SIMPSON

IMPRINT IN HUMANITIES

The humanities endowment
by Sharon Hanley Simpson and
Barclay Simpson honors
MURIEL CARTER HANLEY
whose intellect and sensitivity
have enriched the many lives
that she has touched.

The publisher and the University of California Press Foundation gratefully acknowledge the generous support of the Simpson Imprint in Humanities.

Seek Higher Ground

Seek Higher Ground

THE NATURAL SOLUTION TO OUR URGENT FLOODING CRISIS

TIM PALMER

UNIVERSITY OF CALIFORNIA PRESS

University of California Press
Oakland, California

© 2024 by Tim Palmer

Photographs are by Tim Palmer unless otherwise noted.

Title page photo: The Elk River of Oregon floods into its riparian forest at the height of a winter storm.

Library of Congress Cataloging-in-Publication Data

Names: Palmer, Tim, 1948– author.
Title: Seek higher ground : the natural solution to our urgent flooding crisis / Tim Palmer.
Description: Oakland, California : University of California Press, [2024] | Includes bibliographical references and index.
Identifiers: LCCN 2023031776 | ISBN 9780520382732 (cloth) | ISBN 9780520382749 (ebook)
Subjects: LCSH: Flood damage prevention—United States. | Rivers—United States—Regulation.
Classification: LCC TC530 .P28 2024 | DDC 627/.120973—dc23/eng/20230825
LC record available at https://lccn.loc.gov/2023031776

Manufactured in the United States of America

33 32 31 30 29 28 27 26 25 24
10 9 8 7 6 5 4 3 2 1

Contents

1	The Coming Storm	1
2	The Essential History of High Water	11
3	Rivers Need Floods and Nature Needs Floodplains	52
4	Dams for Flood Control: The Promise and the Reality	71
5	Broken Barriers	92
6	Higher Floods and the Endless Storm	121
7	Floodplains Are for Floods	141
8	The Insurance Connection	166
9	Moving to Higher Ground	194
10	Greenways	243
11	Living with Rivers	259
	Acknowledgments	271
	Sources	273
	Notes	277
	Index	305
	About the Author	311

1 The Coming Storm

Windshield wipers couldn't keep up.

The world outside was blurred, changing, and threatening. I couldn't even see clearly, let alone forecast what was going to happen.

Rain had begun falling in the night. At 3 a.m. I first heard it as faint but steady static on the roof. It grew in my tossing sleep to a staccato drumbeat that kept me awake before my usual 6 a.m. get-up-and-go. I drove to work on a wet road that reflected the gray sky. In low spots I dodged rippling puddles that were quickly filling to the brim, their depths cryptic.

The downpour continued all day. People complained because the spring and summer of 1972 had already been too wet, bad for gardening and ball games, with mildew oozing where you wouldn't expect it, all accompanied by a sour whiff of rot in the woods and suspicion that something was wrong, somehow wrong. Plus, we simply missed the sun in what was feeling like a different world from what we had always known, leading some to wonder if the climate itself was changing.

Few thought of it at the time—I certainly didn't—but a fateful setup, far beyond our reach, had fallen into place. A rainy spring had saturated the soil, a lineup of storms had dampened the first weeks of June, and a prolonged buildup of clouds were all followed by the approach of hurricane-driven air that had earlier swept across the South, causing minor flooding before tracking northeastward and sojourning to sea. Good riddance! But then, from the warming ocean of early summer, the stubborn storm resupplied itself with additional moisture—a lot of it— and unexpectedly doubled back on the mid-Atlantic states with renewed punch, power-packed in ways no one realized or imagined. The soggy atmosphere, swirling in a counterclockwise cyclone filling the weather-map of eight states, no less, bore directly down on central Pennsylvania.

As the hours passed, the radio warned of high water and local drainage problems. Though the full scope of rude reality had not yet sunk in for any of us, my boss had a sixth sense for sniffing out trouble, and he gave us all a greenlight to head home. I had a lot to do, but to play it safe, I grabbed my raincoat and left my job as Lycoming County's environmental planner a bit early.

Though it was only 4 p.m. on the year's longest day, June 21, the cloud-shrouded sunlight dimmed behind a murky overhead brew that just wasn't going away or even budging. The gloom made the hour seem crepuscular, twilight elbowing in ahead of time, cheating us out of the solstice with its normally lingering daylight that conveys leisure, comfort, and celebration of nature's more benign cosmic inevitabilities, not to mention expectations of summer with its smiley themes of carefree youth, hope, and renewal. Instead, runoff gushed from houses' overloaded downspouts and through culverts tunneled under roads in channels churning full and angry, belching with mean backups that could cover all signs of the pavement. The ground had become a saturated sponge capable of absorbing no more. But the water continued to pound down from the sky anyway, and while a single day of hard rain was not uncommon at that moist, temperate latitude—due west of New York City but hidden within rugged folds of the Appalachian Mountains—the veil of moisture and its relentless accumulation augured something strange.

The water's impatient overtopping of ditches, its brown foam sudsing every stream leading to the Susquehanna River, and its insistence on flowing even in usually dry places all spoke of something even darker than the unseasonably dim light. It was only rain, only water. Yet its rise, as I neared my home, hinted at a fate bending off the charts. As the currents of river-sized Pine Creek came into view from the road, their menacing effects

The Hurricane Agnes Flood approaches its crest in Pine Creek, a West Branch Susquehanna tributary, at the village of Cammal, Pennsylvania. This flood was the costliest in American history up to that time, in 1972, with damage occurring across eight states.

seemed to multiply. A lot of runoff is one thing, but the push, the urgency, the whirlpools, the furious violence of something as artistically fluid and normally appealing as water all seemed to dismiss not only any reasonable expectations we might have had, but also any lingering sense of control and security. If any of us had clung to illusions about being in charge of what happened around us or, for that matter, to any wishes regarding our own immediate fate, those ideas were called into question the way a naive sense of immortality might fade when one is being wheeled into the local hospital's ICU.

The Hurricane Agnes Flood of June 1972 would grow to become the most damaging and costly flood in American history up to that time. I happened to live at the epicenter of the storm.

Little did we know, during the first day of rain, that the impending flood would inundate whole cities by surprise, overtop dams no one ever expected to fill, and sever levees that for decades had created a false sense of security and ultimately an expensive fantasy for all who lived behind them. A rainstorm of three days' duration transformed our world by running off the land in currents making a war zone out of the places where we had so peaceably been living. Following the flood, the aftermath in stagnant water would breed mold and disease that overnight turned cherished homes into biological battlefields. With the waters, the emotions and apprehensions grew, ascending toward fear while whole communities plummeted toward chaos, their lines of access and communication cut by a few strokes of a sharpened climatic knife. Beyond the nightmare that lay ahead of us, the opaque depths of every river and stream would lead to extended uncertainties, troubled introspection, and eventually public deliberation and divisive uncertainty about how to deal with the next flood. That uncertainty and divisiveness continue to this day.

My wife and I and my visiting twin nephews—an impressionable nine years old and thoroughly game for what I was successfully spinning to them as "adventure"—schlepped our belongings up to the second floor of the cabin where we lived along Pine Creek, and we all slept in the rustic loft through that second night of rainfall while the storm continued to savagely transform my beloved and tame little waterway, known for its scenery as the "Grand Canyon of Pennsylvania," into a torrent with as much flow as might normally be expected in a whole branch of the arterial Susquehanna, which, 250 miles southward, became the principal source of Chesapeake Bay—largest estuary on the East Coast. Spanning it all, the stage had been set for a flood like never before seen. The drama unfolded hour by hour in the days ahead.

Strangely intrigued, being stranded there, with water bearing down all around, I could feel something inevitable and timeless beyond the imminent destruction and mayhem—something mysteriously suspenseful and undeniably exciting about the raw power of nature. What I felt seemed almost like an appreciation of beauty that would not have been so secondary to a powerful element of dread had we not been standing and living directly in the water's path. But we were, our lives and homes a bull's-eye in the storm's massive, slowly spinning target.

Dawn broke the next morning with continuing rain, and the main road offered temptations to escape, but reports already circulated about its closure. We couldn't get out. Little did we know then that even before the flood's crest, we couldn't have driven up or down the Appalachian valley more than a few miles, owing to mud-sliding hills, undercut banks, overflowing tributaries, and historic iron truss bridges soon to be tipped over by the force of flow, twisted, crumpled like ERECTOR-set toys carelessly stepped on and then—as if kicked across the room—gone.

Reduced in our seclusion to spectator status, my wife and nephews and I suited up in rain gear and walked down to the edge of the flood to watch a half-floating parade of whole mobile homes along with amputated garages, sinking appliances, and waterlogged furniture, all speaking of people's suffering upstream and all now ingloriously battered by floating logs speeding buoyantly on the crest. The unsorted mess—liquid, solid, and a full continuum in between—rolled on waves that looked more like ocean swells than anything ever before seen that far from sea.

People's response in the face of this spill and peril from the sky had begun, as mine did, with carrying treasured belongings up to a second story, provided the flood victims had one. But in many cases a second story was not enough. Other measures—sandbagging doorways, trenching lawns, pumping basements, blaming, regretting, praying by some, and finally, soaked and muddy, evacuating just in the nick of time—were all hopelessly vexed if not fully thwarted by forces far greater than any response we could muster. Expectations of security were crushed. Whatever they might have been, every person's prospects for the future were severely diminished, all in a matter of just days and hours, all by just one storm falling out of the heavens above us.

It occurred to me that many of our decisions about where to live might have been mistakes, grievous ones overlooking a limitation as serious as high water delivered by something as inevitable as rain. I was reminded, somehow from distant memories, of my efforts years before to craft and then place wooden birdhouses for bug-eating swallows outside the home where I lived. In spite of my best offerings, the birds never settled there. Later I learned that an overhead telephone wire, strung within a short flight of the boxes, was close enough for predatory English sparrows to station themselves and successfully target the eggs and chicks in the homes I had intended for the swallows—something the invasive sparrows can't do if they lack a nearby perch for staging their ambush. Fortunately the birds I sought to lure in with my cute little houses harbored a deep evolutionary imperative to go somewhere else. They knew that my choice for their homesite was no place to lay eggs and raise chicks. Regarding floods, I wondered if we had lost that bird-brain instinct for survival. It was nowhere evident.

Quiet retrospection such as this, throughout the days to come, was interrupted by the gusty loud approach of National Guard helicopters in Army-green military modus bringing our isolated and needy Appalachian village food, their whacking rotors sounding a lot like newscasts from horrors of the war that still raged in Vietnam. Cut off from the world, as anachronistic as a cluster of thatched-roofs predating technology but suddenly exposed to napalm, my little burg under postflood conditions might, for all its vulnerability and damage, have reminded you of a scene not unlike that of the Southeast Asian jungle. But there we were, twentieth-century Americans only an afternoon's drive from the glamour and galleries of New York City.

The Hurricane Agnes Flood impressed me as a force of nature that could not be challenged. Since it resulted from only a few days of heavy rainfall, it seemed as though it would inevitably happen again, in one form or another, in many parts of our country. And in fact, it has.

In the pages ahead, I could focus on the hardships of flooding and on people's heroic and personal responses to them, as nearly all journalists do when adopting this topic. Tempting and compelling as those personal narratives are, I'm drawn to the larger picture of flooding in America. What has it meant? How have we responded? What should we anticipate? What have we done wrong? What must we do differently? I'm driven by my own personal memories and by intimate knowledge that difficulties during floods are real. However, as an investigator and reporter here, I'm motivated more by what floods mean to rivers, to nature, to society, to public policy, and to all who have literally or intellectually ventured into the rising waters' path. As challenging as floods have been, they're now becoming more threatening, and in the years ahead they bode greater importance in our lives, economy, and environment.

I'm prompted to tell this story now, a half-century after the clouds of Hurricane Agnes stubbornly darkened my home, because now we know beyond doubt that a warming climate fuels the storms of the future, and that it portends not just more floods like those of the past, but manifestly stronger ones. The changing climate will occur everywhere, and for the rest of time as we know it. The floods of the future will make those irrepressibly rising waters that threatened my cabin along a Susquehanna tributary look exactly like what they were: just one incident, one moment in a timeline that now reaches beyond all horizons to new floods and to new truths, doubts, and resolutions. As we will see, much of what we've done about floods has not been adequate or, for that matter, well-conceived, and now our responses will be decidedly more consequential to our lives, our communities, our rivers, our surroundings, our everything.

To be unaware of what was going to happen back on June 21, 1972, might have been understandable, but to ignore the hazards of our climatic and hydrologic future in 2024, or any forthcoming date, would be inexcusable, in every sense of the word, to ourselves and to the coming generations. Quite simply, we need to rethink and reform our approach to the perennial and intractable problem of flooding.

Back in 1972 I didn't know that federal initiatives and public responses to deal with this most-common and profoundly inevitable form of natural disaster amounted to one costly attempt after another, the two common denominators being extreme expense and frustrating futility. It's not that what we've done has always been a mistake—many efforts have served us well—but too often we've paid a lot to do the wrong things. First came levees that too often got overtopped or breached. Then we built dams to catch runoff, but they often failed to reduce water levels when they mattered most, during catastrophically big floods. Some dams—cracked, leaking, or otherwise weakened—have ended up being more hazardous than the danger they were intended to avert.

Finding that our dam- and levee-building were not adequate, Congress launched a flood insurance effort, worthy if not compelling in concept and foresight but compromised by special interests, marinated in the worst kind of politics, and entangled in bureaucratic complications, ultimately delivering only a token of the intended reform and reduction of damage while ringing up a deficit of $24 billion and counting, as I write this line. The insurance program has become as unsustainable as people's occupation of the low-lying properties the program sought to insure. Worse, it has ended up incentivizing new development on floodplains rather than preventing it—exactly the opposite of what was needed, and directly counter to outcomes desired by all but the most cynical of self-interested power-players. The law of unintended consequences looms large in this story.

After two centuries of attempts to control floods and fifty years after my confrontation with Hurricane Agnes, the damage floods do is now greater than ever. Congress continues to dole out billions of taxpayer dollars for relief on a never-ending cycle: suffer, spend, recover; suffer, spend, recover, repeat, repeat, repeat.

I knew none of this back during the 1972 deluge, and here's something I knew even less about: healthy rivers depend on floods through natural patterns and cycles. If we want fish, and if we want whole river ecosystems to function, including their provision of essential water supplies, we *need* floods, strange as that may sound. Over time I came to realize that the tragedy is not that floods occur, but that we've built homes, businesses, industries, roads, railroads, and even hospitals directly in harm's way.

Floods happen frequently and with force that ranks them among the most impressive of natural phenomena, right up there with the geologic extravaganzas of earthquakes and volcanoes. Momentous climatic events of our time also include hurricanes, whose coastal winds and battering surfs tend to overshadow, in the media, the intense rainfall delivered by those same storms inland from the coast and causing rivers to overflow. To the mix of these graphic geophysical dramas we can now add drought, which is ironically the opposite of flooding, and we must also add the subsequent western wildfires, which in recent years have raged with the searing fallout of a climate that's desiccating whole regions—a continental blow-dryer with no "off" button, entirely out of control, hardwired to continue. But floods are the most common of all these disaster phenomena, and they cause some of the most regrettable human losses, elicit some of the most remarkable responses, and result in some of the most concerted public efforts in prevention and aid, if not the needed avoidance.

For me, the high water of 1972 was a formative and foreboding introduction to floods. It stretched my perception of the damage that can occur and of the myriad and profound ways that floods will influence our future, whether we live in the wet zone, as I did, or not.

While we've stumbled along in our reactions to floods for two centuries, the consequences have now caught up. Still, the most common response to future floods' inevitability and the need for us to get out of the way might best be described in one word: *denial*. But now, because of a warming climate, the floods of the future are rising. They're becoming increasingly frequent, more intense, longer lasting. The age of denial needs to be over.

Let me note, here at the outset of my story, two disclaimers. First, while river flooding and coastal flooding both rank as ominous problems, this book is principally about flooding by rivers. Many of the issues are similar, but flooding along our ocean edges involves hurricane winds, terrifying waves breaking ashore, and rising sea levels, all of which have recently been addressed in other fine books. However, the coastal storms also bring drenching rainfall to entire river basins, and so I draw on relevant experience and data from the coastal regions and their hurricanes whenever needed in order to understand the big picture.

Second, though critical of what we've done, the following chapters are not an argument that building dams and levees for flood control was unconditionally a mistake. The structures have reduced flood damages in many cases. I do not suggest that all those efforts have been wasted or undertaken in bad faith. To the contrary, most decision-makers at the time did what they honestly thought was best. My point, rather, is that approaches of the past have proven inadequate. We've avoided alternative strategies that now must be pursued with more vigor, and the warming climate demands that we work more, better, and faster to address the problems that confront us. We've reached what must become a turning point. From this day forward, the changing climate with its rapidly rising waters will give us no choice but to regard floods differently.

As the stream in front of my house rose and then fell, the flood made me curious, and then it compelled me to learn more, and more, at a deeper and deeper level. As I became aware of our failing response to high water and of the cycles of life and economy surrounding it, I was drawn irresistibly into the currents. Some of these were both literally and figuratively clear as crystal, luring me along, while others swirled muddy and chaotic, challenging any form of navigation or transparency—physical, emotional, or intellectual.

I struggled to escape the classic trap that plagues the researcher, if not everyone, in so many endeavors: the easiest person to fool is yourself. Assumptions and predispositions can lead us dangerously astray. So I left

In January 2023 a series of storms caused flooding along the entire length of California. Here the Russian River and its tributary, Fife Creek, have backed water into a low-lying neighborhood of Guerneville.

my own past and beliefs behind in order to see anew and to grasp what was real, consequential, and important in all perspectives of the picture that slowly came into focus. Together, those currents of curiosity led me to every problem of flooding that I encountered and to every solution that came to light. On that rainy day in 1972 I couldn't even see clearly, let alone forecast what was going to happen to my river, my home, my job, and my life. But now I know.

In the murky floodwaters lies the clarity of an essential lesson, important to all: as a society, we've regarded the laws of nature as optional when in fact they are absolute. Meanwhile we regard our own habits, customs, and laws as immutable when in fact we can change them whenever we collectively decide to do so. In our minds we have perfectly reversed the way the world actually works, and nowhere is this delusion more evident than along a flooding river. Perhaps nowhere does this lesson make so much difference. For two hundred years, that troubling truth has accompanied the history of flooding in America.

2 The Essential History of High Water

UNKNOWN AND UNEXPECTED

After two decades of searing drought, no one likely expected anything different, but then, beginning on a chilled and drenched Christmas Eve, 1861, intense rainstorms whipped across the West Coast. For forty-three days water fell from the sky with scarcely a break. Engorged rivers pounding down from the Sierra Nevada and from the Coast Ranges overflowed and submerged California's Central Valley lowlands, creating an inland sea 300 miles long and 20 miles wide, two-thirds the size of Lake Erie, saturating a landscape later nicknamed the Golden State in recognition of its famously cloud-free sunsets and warm glow of seasonally arid grasslands. Today's alignment of Interstate 5 became boatable for hundreds of miles, not to mention bank-topping flows caused by the same storms in Oregon, Nevada, and Arizona.[1]

Thousands of people in California died as the flood destroyed one home in eight. Ten feet of water flowed down Sacramento's streets, drowned living rooms, and soaked second-story bedrooms. A shoddily built levee failed and, turning against its creators, trapped runoff *inside* its barrier, a filthy bathtub of stagnant brew lingering days, weeks, months. After a chain-gang

TIMELINE OF MAJOR FLOODING EVENTS

1200	Cahokia flood, Mississippi River
1650	California flood in sediment record
1850	Charles Ellet flood-control plan for the Mississippi River
1861	General Humphreys plan for Mississippi levees
1862	California megaflood
1889	Williamsport flood, Pennsylvania
1889	Johnstown flood, Pennsylvania
1917	first Flood Control Act by Congress
1927	lower Mississippi flood
1928	updated Flood Control Act
1928	St. Francis Dam failure, California
1936	Pittsburgh flood and Flood Control Act authorizing studies for dams
1937	Ohio River basin flood
1938	Los Angeles River flood
1955	Northern California flood
1964	Northern California flood
1972	Hurricane Agnes Flood, Pennsylvania and mid-Atlantic region
1972	Rapid City flood
1972	Buffalo Creek Dam failures, West Virginia
1976	Teton Dam failure, Idaho
1977	second great Johnstown flood
1983	Glen Canyon Dam crisis, Arizona

1986	Sacramento flood
1993	Great flood of the Mississippi
1997	Sacramento flood
2005	Hurricane Katrina Flood, New Orleans
2008	Mississippi River flood
2010	Nashville flood
2011	Mississippi River flood
2014	Elwha River dams removed, Washington
2015	Hurricane Patricia in the eastern Pacific Ocean, strongest hurricane ever recorded
2016	Greenbrier River flood, West Virginia
2017	Hurricane Harvey Flood, Houston, heaviest rainstorm in American history
2017	Oroville Dam crisis, California
2018	Hurricane Florence Flood, Lumberton, North Carolina
2019	Mississippi basin floods, which continued for a record-breaking nine months
2020	Tittabawassee River dam failures, Michigan
2020	floods in the East and Mississippi basin
2021	Nashville flood
2021	Hurricane Ida and New York City flash floods
2022	floods in Yellowstone National Park, Kentucky, Illinois, St. Louis, and Florida
2022	Hurricane Ian in Florida
2023	January floods throughout California and record snowfall and runoff in March

breached the levee's low spot, water drained partway but still pooled head-deep on many streets. Leland Stanford accepted inauguration to his brief term as governor while floodwaters flowed. The chief executive then returned by rowboat to his home and ingloriously entered through a second-story window. State government fled to San Francisco, facing bankruptcy while the nation descended into the darkest depths of the Civil War.

Though only a century and a half ago, and unlike the lesser disaster of San Francisco's infamous 1906 earthquake, the 1861–62 high-water-apocalypse was all but forgotten. Even the cyclopedic *California Water Atlas* dismissed the flood to sidebar status because "records of this event are too incomplete for a certain assessment."[2] For many people, California's landmark flood has not been adequately factored into floodplain expectations of the modern era, replete as it is with engineering prowess and GIS data. Such societal amnesia could have dire consequences for essential preparedness.

Yet, history can inform and warn us about the future. So let's consider, in this chapter, the nature of flooding in America and recall some of the most important floods, with a special eye for those that resulted in changes to our attitudes about flooding, our responses to it, our laws regarding it, and our efforts to address the problems that floods cause and also the solutions that we have and have not pursued.

Since 1862 the population of California's flood-prone Central Valley, not including the rest of the state, has exploded from 61,000 to 7,000,000. With this meteoric growth, a similar flood would be "incredibly more devastating," according to hydrologist Michael Dettinger of the US Geological Survey and geologist B. Lynn Ingram of the University of California. They took the issue of California flood potential to a new level in the *Scientific American* article "The Coming Megafloods."[3]

Not only is the vulnerable population of the Central Valley now 115 times greater than it was then, but some of the real estate has subsided. Not in market value, which has skyrocketed, but *physically* subsided, sunken as much as 30 feet, owing to uncontrolled groundwater pumping by Central Valley farmers who for a century refused to be constrained. In places the depression is enough to totally submerge a three-story home that might have escaped flooding in Governor Stanford's day. But even back then, without surface subsidence, widespread losses were tabulated as "total."

Low-lying wetlands and floodplains that settlers avoided in the 1800s because of permanent surface water, frequent floods, and whining clouds of bloodthirsty mosquitoes have long since been developed with farms, houses, infrastructure, and cities. Flood-control efforts have included dams, levees, and even a channel excavated as a man-made river by the Army Corps of Engineers in the early 1900s to allow floodwater easier drainage and escape into San Francisco Bay. But exceeding the pace of those improvements, powerplants, freeways, airports, industries, and even nerve centers of emergency responders, not to mention hundreds of thousands of homes unimagined a century ago, have been built on flood-prone real estate. People's survival will now depend on the uncertain security—or one might more accurately say the proven insecurity—of hundreds of miles of Central Valley levees when the next mega-storm arrives.

What makes rediscovering these historic facts about flooding in the distant past so relevant today is that those truths have not been factored into many of the modern projections and practical applications of land-use planning. That's bad, because new revelations dwarf our concept of what the largest floods were and, more important, what they inevitably will be, and not necessarily that far into the future.

Even though the flood of 1862 was documented by witnesses at the time, American history, defined as the written record, goes back only so far, and not *very* far, specifically to the first notes jotted down by European settlers and Spanish missionaries displacing Native Americans who had kept oral rather than written accounts of the past. These, of course, went back much further, for anyone inclined to listen. But tools of paleogeologists now enable analysis of silt deposits to determine the size of floods that occurred before anyone penned a journal entry or newspaper dispatch. All this constitutes what we might call "deep" history of the subject central to this book.

For a window to that mysterious past, geochronologists examined sediment along calm waters in West Coast rivers and estuaries. Each deluge, it turns out, delivers a signature quantity of suspended silt that settles when the currents eddy out and stop moving, and it lingers there as if frozen. The depths of those deposits, measured in lab samples, reveal the magnitudes of floods from long ago, much as dendritic rings in trees indicate climatic conditions when ancient trees, such as 5,000-year-old bristlecone pines, were young.

Using the silt-survey technique at strategic locations in San Francisco Bay, scientists have unearthed the fascinating past of epochal floods predating written records by whole millennia. Chronicles reliably written in the mud indicate that mind-boggling floods have occurred every few hundred years, or uncomfortably less, and that the staggering levels dwarf anything so far considered a 100- or even 500-year flood, which are the metrics that underpin the mapping, planning, land-use regulations, investments, insurance calculations, everyday attitudes, and uncertain comforts regarding security today.

Even for those who know about it, the 1862 flood is often regarded as a freak of the past, unlikely as another Battle of Gettysburg, which occurred not long after. But the new sediment-core samples prove just the opposite. Similar floods—no, actually much larger ones—have reoccurred repeatedly, spilling water across the Golden State's landscape like a slow, dependable, liquid-pumping heartbeat patiently synchronized to geologic timelines of the Central Valley.

Yet a convenient sense of denial pervades and a collective mental block against the apocalyptic surfaces when danger fails to tap us on the shoulder regularly. The comforting denial says, "That can't happen again." Therefore, "We don't need to plan for that." Or, worse, we humans harbor an underlying, unspoken attitude that, for many, is tantamount to an embarrassing belief: "If it hasn't happened to me, it hasn't happened."

But the geochronologists' no-nonsense sediment cores of fossilized mud indicate that, prior to 1862, California megafloods had occurred in roughly the years 1100 and 1400 and then again in 1650, when silt deposits registered a deluge 50 percent greater than any of the others. For a challenging if not mind-blowing perspective here, consider that the notorious 1964 flood of Northern California—widely regarded as the epic high-water climax of modern times or, mistakenly, of all history—left silt deposits in San Francisco Bay of 0.08 inches. But megafloods of the deeper past each left silt layers over 1 inch deep—a staggering twelve times the amount produced in 1964, which gives pause even when you assume some large margin of error or factor in every single one of the flood-ameliorating effects of every modern-day dam, levee, diversion, bypass, and hydrologic strategy for dealing with the problem. To tease out the buffering effect of today's flood-control system raises a vexing veil of probabilities and assumptions but does not undermine a singular truth: those floods of the deep past were big. *So* big that most people don't even want to know about them, let alone think about them.

Climatologists now credit the reoccurring megafloods to "atmospheric rivers" of intensely wet air traveling a mile up in the sky as impressive vapor streams that can be 620 miles wide, 1,200 miles long, and nearly 2 miles deep trailing across the Pacific and aimed like a bloated fire hose directly at California for days or weeks on end. This phenomenon had for years been casually called the "pineapple express" because many of the famously saturating tempests came from the direction of Hawaii. Considering the amount of rain these storms deliver, they might be loosely regarded as a West Coast version of the hurricanes that dump famously prodigious amounts of water on East Coast and Gulf of Mexico states, though hurricanes are better known because of their ripping velocities of wind and photogenic mountains of surf that either remove roofs overhead or undermine foundations underfoot, often doing both, leaving in their wakes citywide junkyards of debris in places such as Galveston, Fort Myers, and even Staten Island. But the less-windy atmospheric rivers account for 30–50 percent of California's annual moisture. In other words, much of the precipitation in the far west comes from relatively few storms each year, which is a meteorological setup for floods.

Now suddenly in widespread use, the phrase *atmospheric river* was coined in 1998 by geoscientist Yong Zhu and meteorologist Reginald Newall at the Massachusetts Institute of Technology when they noticed the impressive climatic features worldwide. They found that beyond the subtropics, much of the world's vapor transport occurred in these relatively narrow atmospheric bands moving west-to-east across middle-northern latitudes, causing not only the California floods but also drenching rainstorms in the East.[4] Minor atmospheric rivers happen almost yearly and account for the typical whopper storm dumping 5 feet of glistening snow on California's Sierra Nevada high country. But a large event can carry as much water as the Amazon, or more. Imagine such a thing, *in the sky!* Then imagine it as intense rain falling for weeks and as snowstorms that close down even ski resorts because chairlifts at treetop heights become mired in drifting snow. I'm personally reminded of a storm-endowed winter when I lived alone in a friend's cabin perched in California's Donner Pass. Daily I was thrilled to ski into spectacular backcountry right outside my door, but I was also destined to spend whole days shoveling my entombed van out from under 6-foot pileups, variously airy powder and leaden slush on that snowy year.

The new data about prehistoric storms tell us something we'd do well to learn: there's no reason to assume that extraordinary floods from atmospheric rivers will not come again. In fact, California may be due for a big one. *The* big one. Dettinger and Ingram wrote that a megaflood has occurred on average about once every 200 to 300 years. Though record-breaking snow and rain in 2023 gave us a hint of what to expect, it has now been more than 160 years since the last monumentally big flood. So do the math, including cautionary margins of error, because an epic flood could come sooner. In a separate analysis where global warming was factored in, climate scientist Daniel Swain at UCLA's Institute of the Environment and Sustainability forecast a fifty-fifty chance of a megastorm sequence of the 1862 scale by 2060, with clear possibilities that it will come before that.[5]

The reach of historical records provides a snapshot of only the past 400 years—barely a blip in the grand scale of climate. Not just in California, but across the country, both flood histories and predictions have failed to acknowledge the largest floods from the distant past, though similar events are certain to reoccur. Consider the Mississippi valley, where in the year 1200 or so, at the Indian site of Cahokia—think St. Louis—7.5 inches of silt was deposited by a flood that would have topped the charts, had there been any.[6]

Consider, even, the Great American Desert. Drawing on evidence from carbon-dated silt deposits, geomorphologist Victor Baker of the University of Arizona estimated that a Colorado River flood of 500-year frequency at the trendy red-rock town of Moab, Utah, would peak with 246,000 cubic feet per second—not just larger, but *two times* anything that was foreseen by engineers who designed the dams now standing amidstream and obligated by our public institutions to contain Colorado River overflows pushed sky-high in the wake of storms. Baker called the dam planners' standards—meticulously calculated by competent engineers but based on adolescent data barely a century old—"completely inadequate."[7]

More on Colorado River surprises later, but back to 1862 and California: when scientists at the Geological Survey modeled a flood like that West Coast, Civil War–vintage episode, but lasting only twenty-three days (*half* the 1862 storm's duration), they determined it could require evacuation of 1.5 million people.[8] Multiple levee failures could lead to a Hurricane Katrina disaster not only in California's capital, but for a whole chain of twenty-first-century cities now lining the state's Central Valley.

Updating this starkly terrifying analysis to 2019 dollars, and focusing strictly on the foreseen economic train-wreck, Dr. Swain estimated that damages would be $900 billion, not counting secondary costs such as deflated tourism—a backbone of the state's modern economy. Fifty levees could be breached and multiple dams could fail. Whittier Narrows Dam on the San Gabriel River alone could endanger 1 million people, a sobering prospect of mayhem that springs to mind when standing on the crest of that 56-foot-tall structure and gazing out across the schematic-like panorama of streets and roofs filling the Los Angeles Basin directly below.[9] Considering new data on climate change, Dr. Swain predicted that what was recently considered a 200-year storm is more likely to occur once per 40–50 years. In other words, events previously considered improbable in two lifetimes could likely occur twice to each of us, rudely interrupting the "if it didn't happen to me, it didn't happen" brand of denial. Elaborate systems of flood-control dams and levees will be utterly overwhelmed by such floods no matter what predictive model is chosen. Speaking of which, a new US Geological Survey analysis of projected storm potentials is underway as of this writing, and the forecast calls for even greater flooding.

The bottom line here, especially difficult for both advocates of structural flood control and real estate developers to swallow, is that our flood-control strategies of the past, relying on dams and levees, will ultimately not work. They'll fail precisely when we need them the most, like a bank defaulting on your savings account the day you're laid off work. Maybe even the day you're laid off *and* the day that you check into that ICU unit mentioned in chapter 1.

Duly warned, we now know that the next 8.6 earthquake in California will not be the only surprise that should not be a surprise. While media have long characterized earthquakes as the iconic California disasters ever since the San Francisco shakedown and citywide blaze of 1906, more than half the state's eighty-five disaster declarations since the 1800s have not been for seismic rattlings and broken dishes, but rather for soggy floods and saturated mattresses.[10] Damage estimates for a big atmospheric-river flood, virtually assured in the future, totaled three times those forecast for a "big one" magnitude 7.8 tremor, rocking our collective boat. So let's set the disaster-mythology straight, here: flooding was the largest disaster event of the past, and it's the largest on the horizon.

New scientific disciplines have uncovered this unwelcome history, but taking the appropriate action requires more than just knowing about floods of the past and the likelihood of their reenactment. As California's quintessential novelist, John Steinbeck, wrote in *East of Eden,* "It is one of the triumphs of the human that he can know a thing and still not believe it."

History provides us with much of what we need to know, but like tragic Steinbeckian characters, we refuse to believe the facts in front of us, and as a result, no one is ready for the floods of the future.

THE ANATOMY OF FLOODING

To help us prepare for the inevitable floods of tomorrow, let's consider the tragic, repetitive pattern of floods in the past. We've suffered through high water again and again since the first New England towns were cobbled together above the banks of rushing rivers, which on fair-weather days provided for water, power, and docking of boats venturing in from the ocean. Those riverfronts appealed as places to live, manufacture, and do business, all given the assumption that floods of the past would not come again.

But they did. Time and time again people have looked out at furious brown waters foaming toward them and felt a churn in their stomachs sourced in something more threatening than anything they'd ever seen. As a printed record of this emotion, a brittle old newspaper caught my eye when, as a teenager, I helped my parents clean out my grandmother's belongings, which spanned a half century in Rochester, Pennsylvania, along the Ohio River 30 miles downstream from Pittsburgh. Hidden among personal memorabilia, a twenty-page photo insert of *The Pittsburgh Press,* titled "The Great Flood," bore the date April 8, 1936. Being a history buff even at a young age, I saved the yellowed photo documentary. The only other newspapers filed deep in Mildred and James Palmer's dusty attic covered the declarations and ends of two World Wars, the first of which nearly claimed my grandfather's life on the front lines in France. Taking an indirect hint from my grandmother's selection of archives, I concluded that floods have been a notable part of American culture—and hardship—going way back.

For all their importance in history and environment, rivers and streams occupy only one-half of one percent of the Earth's surface.[11] Intimately related to this, floodplains that those waterways inundate at peak flows cover 7 percent or so of the US. Both numbers pale in comparison to the role that water and floods play in essential workings of the natural world and also to the importance of flooding on infrastructure, economy, and society. Given the risks, what we *do* on that flood-prone land may be far more consequential than it is on any other 7-out-of-100 acres nationwide. The history of flooding bears this out.

Likewise, the duration of floods represents just a statistical blip on the calendar. Rivers might typically reach "flood stage," meaning bank-full flows, for only a few days every few years or even much longer intervals. But in many ways these are the days that count the most—the ones when earth is literally moved, floodplains nourished, and damage inflicted on whatever happens to be in harm's way. With floods, and on floodplains, you might say that nature gets its biggest bang for the buck. Floods mean that long interludes of lazy flow, uncompromised beauty, and uninterrupted pleasure at the waterfront are punctuated with episodes of what might be termed hydrologic madness. The floods' short durations may account for our collective ability to forget the hazard until it reoccurs, and that amnesia may fester at the root of our inability to address what floods, in fact, require us to do.

"History, in the end, is homage," wrote Professor Douglas Brinkley, author of *The Great Deluge: Hurricane Katrina, New Orleans, and the Mississippi Gulf Coast*, along with other epic American accounts. "It's about caring enough to set the record straight even if reliving the past is painful or disappointing."[12] The history of floods is certainly painful, and the outcomes of our efforts to deal with floods are disappointing if not shocking, frightening, and heartbreaking. But a constructive outlook on a natural phenomenon that will never end—and, in fact, that will intensify and become increasingly prominent—is essential if individuals, society, and government are to cope with portentous forces that grow ever more powerful. Flooding affects not only those who live on the 7 percent of the land that the rising waters will soak, but all of us.

For deep historical perspective here, remember that flooding did not start when we became vulnerable to the damages, but rather dates to the origins of the hydrologic cycle or, in other words, to the early days of water on our planet. With a bit of poetic license and apologies to those aware of the Hadean Eon, we might even say that floods date to the beginnings of time. They hold key importance in any creation story about the Earth and the life upon it.

So let's step back for a moment beyond history and into the realm of natural history, and recognize, for starters, that floods are not all the same. They differ, of course, in their intensity and duration and also in their origins, depending on weather, geography, and human manipulations of flow and, now, of climate.

At the opening of this chapter we saw one type of flood rising from atmospheric rivers pounding the West Coast. For another, consider summertime air superwarmed and thereby capable of transporting a lot of moisture over East Coast states where the Appalachian Mountains bulge upward and cause wind to push clouds higher, which cools them until vapor condenses and falls as rain. Fertile storms can stall and pummel the earth all day and all night with double-digit accumulations. Streams then spike as flash floods with explosive crests giving little warning when thundering through narrow valleys. Flash floods killed 100 people in 2021 alone and rank as the most dangerous type of naturally occurring high-water event (the *un*natural is something else; we'll get to dam failures later).

One of these flash-flood storms occurred not long ago, on the night of June 23, 2016, over the Greenbrier River basin in West Virginia. A muggy cloud-covered afternoon eased into a hot sultry evening. Breezes of damp air accelerated to erratic gusts, and a gray underbelly of nimbostratus clouds corrugated the sky like a lot of sponge blotters in surreal blobs that looked low enough to scrape mountaintops, which they did. Supersaturated air hovered, scarcely moving at all, and the rain being squeezed from those clouds began to drench the Greenbrier, then increased without letup. The impressive downpour doubled and tripled in strength as thunderstorm soundtracks cracked and boomed from a black sky hiding every one of the

5,000 stars that might otherwise have dotted the night. Water fell with deafening force on rooftops, making it impossible to talk and be heard inside. I happened to be sheltered in a cabin perched above the New River Gorge at the time, not far from the insanely beating heart of that infamous storm. I planned, of all things, to paddle on whitewater of the New the next day—a trip we cancelled.

Twelve inches fell in twelve tumultuous hours. The sheetwash—meaning water that pooled on the ground's surface, all around—lay an inch deep at times, even in places where you'd expect rain to run off instantly. But as fast as gravity could work, all that water headed for the lowest spot it could find, joining numberless other flows and surging together into ankle-deep creeks that bloated to 6-foot depths in an hour. One stream tragically swept away a four-year-old boy. Dawn broke with the town of Marlinton transformed to an inland sea, streets undercut, cars crumpled by currents, highway connections severed, residents mourning their losses.

Downstream from its confluence with the Greenbrier, the arterial New River swelled from a hefty 8,000 cubic feet per second—itself a muddy summertime surprise that brings to mind the beefy Colorado River in the Grand Canyon—to 100,000 cfs after an overnight infusion unimaginable to anyone who thinks that nature might be predictable, constant, or controllable.

Twenty-four people died. Water pinning drivers and passengers in cars became the chief terror, as it typically is, accounting for 64 percent of all flood deaths.[13] It turns out that driving a car is the most dangerous thing we do during floods, outpacing by far people clinging to rooftops in classic photos that make the news.

The National Weather Service reported that a similar storm was likely to reoccur only once every 1,000 years. However, challenging such math, a similar Greenbrier flood had occurred just 21 years before, killing forty-seven people in 1995, and again only one year after that, when President Clinton declared West Virginia a disaster area four separate times, like listening over and over to a song you really don't like. Major floods strike on average once every eighteen months in this state where tourist-bureau signs welcome visitors to "Wild Wonderful West Virginia." One might imagine a range of alternative slogans based on local hydrology and our responses to it.

Another distinctive type of flood comes not in summer, but with heavy rain in winter or spring. Think of pouring a teakettle of hot water onto a patch of snow; you instantly get both as runoff. The "rain-on-snow" event is among the most common, destructive, and repeated sources of epic floods because the melting snow doubles, triples, or quadruples the impact that rain, alone, would impart. The snow, in effect, is programmed for delayed runoff, but that program is cut shockingly short in this scenario, and the accumulated precipitation of days, weeks, or months flushes at once. We saw this in March 2023 when fresh snow 15 feet deep in the mountains above Los Angeles was melted by an infusion of warm water from an atmospheric river pounding the West Coast.

Another class of flood goes hand-in-hand with hurricanes, which are the most frightening of climatological events, as anyone will attest who has cowered inside during storms that can shatter windows and worse. Personally, I had never considered dying on the Gulf Coast of Texas, but during a tropical storm there once, I had to wonder. These summer and autumn events begin when the surface of the ocean, between latitudes of 5 degrees (near the equator) and 30 degrees (northern Florida), warms to 80 F, the tipping point when water heats the atmosphere above it enough to cause exceedingly damp air to rise and create a low-pressure vacuum that sucks in adjacent air. Supersaturated updrafts accelerate and climb as high as the continent's tallest mountain, 20,000 feet. Meanwhile the Coriolis effect—a complicated deflection of wind caused by both the shape and turning of the Earth—causes the entire mass of agitated, low-pressure air to rotate counterclockwise in the Northern Hemisphere. Expanding to 300-to-600-mile diameters, this circular fury migrates as a unit, like an old-fashioned wooden top spinning slowly across the floor and aiming northeastward with prevailing winds in a route roughly paralleling the Atlantic coast.[14]

Best known for their roof-ripping blow and window-shattering surf that together batter everything near sea level, hurricanes also push their clouds hundreds of miles landward, and heavy rain can spread completely across the Appalachian Mountains and north through New England. Extraordinary amounts of water drop when the prevailing winds stall, leaving the storm to rain in-place for days at a time, causing some of our largest floods—Hurricane Agnes, for example. The warmer the tropical air becomes, the more intense the hurricane, and in the age of global warming caused by our burning of fossil fuels, you can tell where this scenario is headed.

Even more explicitly man-made are the floods that descend as a wall of water when upstream dams fail. Remember the accumulated runoff of rain-on-snow? Well, dam failures accentuate that one-two punch to horrifying extremes. Compressing many storms and even several years of accumulated runoff into one catastrophic burst, floods from ruptured dams don't creep upward inches at a time the way that even the leviathan Mississippi might do during the soggiest springtime. Instead, they arrive like tsunamis with destructive forces more concentrated than nature could ever dish out.

Similar to dam failures with their startling speed and crushing volume, yet another type of flood disaster is caused by mudslides and debris flows, which avalanche when heavy rains pound steep slopes and destabilize them.[15] The problem becomes extreme where wildfires have stripped protective vegetation from mountainsides, especially in the West's biome of shrubby chaparral. The fires torch plant cover, along with shallow roots that otherwise stabilize soil, and that alone is enough to spike runoff. But multiplying the hydrologic threat, the leaves of many chaparral species contain water-repellant oils as an anti-evaporative defense against drought. After fire volatizes those oils, some of their smoky residue sinks to the ground, infiltrates, and encounters cooler soil several inches beneath the porous surface soil. On contact with that subterranean chill, the gasified oils abruptly recondense to form a waxy, membrane-like deposit that repels further percolation of rainfall. Picture a plastic tarp spread out on a whole mountainside with a few inches of loose soil on top. When rain comes, water quickly saturates that shallow veneer of soil sitting on the oily membrane, causing the soil to liquify, shear, and plunge catastrophically downhill. The earth-and-water slurry gains speed; plucks rocks, boulders, cars, and houses from its leading edge; and rolls them together like lava in a homogenizing tumbler. Notorious in Southern California, this shocking chain-reaction of climatological, biological, and hydrological phenomena is beginning to occur elsewhere as the climate warms, forests dry out, and chaparral with its volatile oily foliage spreads.[16]

Even a cursory recap of these types of floods throughout American history could fill a book fatter than the one you now hold. So, in the sections of this chapter that follow, I'll highlight a select group of floods that are particularly symbolic of their times, that triggered shifts in our awareness of high water along with our commitments to respond, and that underscore our awareness of the science, policies, and politics enmeshed in these events that are at once profoundly natural and tragically unnatural. The essence of our response, with its complications across a broad canvas of culture and environment, will surface in later chapters, but for starters, consider a few of America's landmark floods principally in the way that they were regarded during their times: disasters causing widespread loss and suffering.

A LEGACY OF HIGH WATER

In the summer of 1889 torrents rained on Appalachian mountains that had been stripped of protective vegetation and scarred by subsequent runoff during the lawless, cut-everything logging boom of the nineteenth-century Northeast. Without tree cover that had earlier shielded the rugged countryside and absorbed the soaking seasonal storms, Williamsport, Pennsylvania, was inundated on June 1 when the West Branch of the Susquehanna River drowned homes up to their ceilings.

Here in that era's timber capital of the world, millions of logs, corralled at riverbank sawmills, were swept downstream to become battering rams the whole way to Chesapeake Bay. A reporter stationed in the city wrote, "The loss is awful. There is mourning everywhere for the dead. The mayor sends this message: Send help at once, in the name of God, at once. There are hundreds utterly destitute."[17]

Even though flood-prone land constituted only single-digit percentages of acreage nationwide, waterfronts like those of Williamsport had been universally targeted for development. As a result, horrific losses were repeated at hundreds of cities, towns, and rural outposts when the nation transformed from agrarian to industrial and urban. While brown waters receded, farmers pried bloated carcasses of dead animals back into subsiding currents, road crews scraped flood debris off to the side, and homeowners shoveled mud from every room in their houses.

After escaping, or not, from threatening epidemics of typhoid and cholera broadcast by the lingering cocktail of runoff blackened by back-flushed sewage, people's memories of terror were soon lost in the rush to recover, and with a large dose of denial, flood victims rebuilt. Meanwhile, any role of government to limit development in hazard zones or even to warn people about floods failed to appear on public agendas. A spirit of indomitable expansionism fueled an entire nation rapt with the gospel of growth and progress, and floods were not going to get in the way. Underpinning this confident vision lay a baseless hope that floods would not happen again— at least not in one's lifetime. Floods were likely forgotten until hearing the next warning to evacuate. But most flood victims heard no warnings at all.

JOHNSTOWN

For sheer violence and tragedy, nothing matched the 1889 flood at Johnstown, a small industrial city crowding streamfronts and fingering up hillsides in western Pennsylvania.

Long before any progressive era of regulation for public safety, a dam on South Fork Run, feeding the Little Conemaugh and then Conemaugh River, was shoddily built 72 feet high in order to feed water to a cross-state canal company. Railroads drove that canal system to default before it opened, and the purposeless dam was sold, then resold to a clique of Pittsburgh industrialists who adopted the reservoir as a private resort and yet somehow regarded its maintenance as someone else's problem. Houses around the artificial lake, of course, sat above the level of the dam and would not be threatened by its failure. The structure was sloppily repaired, its spillway neglected, its overflow culverts blocked even after repeated warnings.[18]

The same torrential storm that flooded Williamsport saturated mountainsides upstream of Johnstown on May 31. Runoff filled the impoundment and then overflowed, rupturing the dam, which released a 40-foot wall of water that tore downstream 50 miles per hour, pushing on its tsunami everything it encountered. More than 2,200 people perished and another 970 went missing as the flood splintered homes by the hundreds and tumbled even railroad cars into the flow. Floating debris and whole houses lodged against a stone bridge, which held long enough for flames from residents' still-glowing lanterns to ignite the combustible suspension of homes imbricated against each other by the water's force. The inferno torched everything above the waterline, burning for three days after engulfing eighty victims who had miraculously survived, until then, clinging to floating rooftops.

Even though the disaster was as human-caused as was the toppling of New York's World Trade Center 112 years later, the flood was judicially termed an "act of God." Courts never held the private club of industrialists who owned and neglected the dam accountable, and for decades our governments did nothing to prevent such disasters from reoccurring.[19] Neither the Johnstown flood nor others were much of an impetus for the states, the federal government, or people at community levels to do much about the hazards of flooding. Nothing, that is, until the accumulating development in danger zones, combined with reoccurring floods, together made what had been a troubling and local problem chronically worse and widespread.

SUFFERING THROUGH STORMS

Confronted again and again by losses and tragedy, American sentiments evolved from a sense of innocent victimization, willful ignorance, and repeated denial and grew through painful repetition toward a political consensus that something must be done about flooding. Even then, in keeping with the hubris of the times and of the frontier mentality that persisted long after historian Frederick Jackson Turner in 1893 declared the frontier gone, people didn't do much to avoid floods. What they considered, rather, was what they could do to stop the floods from occurring. They opted, in other words, to try to control nature rather than temper the actions of themselves and their neighbors.

This attitude prevailed even when people confronted the Mississippi's overwhelming force of flow. Third-largest basin on Earth, the heartland artery draws water from 40 percent of the nation outside Alaska. The main stem receives the Ohio River—surprisingly twice as large in volume where the two mega-streams meet—as well as the Missouri River, which stretches longer than the Mississippi but drains drier territory. Written accounts of the river's legendary floods go back to 1543 and the brutal conquistador, Hernando de Soto, who noted a forty-day flood stalling Spanish explorations at today's site of Memphis.

While floods routinely overtopped the banks of the great river, nothing beyond the construction of local levees had been done to challenge the flooding pattern for 300 years after the Spanish first arrived. But to consider the options, Congress commissioned engineer Charles S. Ellet to prepare America's first river management plan. His 1851 report urged that floodplains be set aside for high-water overflow and storage—a prescient recommendation ignored by one and all. The comprehensive Ellet also advised the government to respond with dams, levees, bypass channels, and what would later be termed watershed management.[20] Informed better than anyone of his day, he warned that relying on levees was "a delusive hope, and most dangerous to indulge because it encourages a false security."[21]

No matter; in what became a pattern, Congress adopted a sharply contrasting plan by General Andrew Humphreys of the Army Corps of Engineers, who in 1861 recommended that the Father of Waters be completely contained by levees, walled off in a channel divorced from its floodplain, with unsubstantiated hopes that the river's natural overflow systems, of millions of acres of floodable lowlands and thousands of miles of back-channels carrying high water, were superfluous to the river's function of water transport and that the floodplains of the ages would somehow not be needed for the floods that continued to come.

In 1912 and 1913 high water crests of the Ohio River spilled onward to the Mississippi and exposed the inadequacy of levee defenses built by local districts. As part of an investigation ordered by President Wilson, the Corps doubled down on levees and concluded that America needed stronger federal programs for flood control. Though Humphreys's approach favoring levees prevailed, the Corps also endorsed the idea of moving valuable property off the floodplains—a farsighted recommendation that failed to see the light of day.[22]

Finally motivated to pass a bill, Congress in 1917 adopted its first Flood Control Act, exclusively addressing the lower Mississippi and Sacramento Rivers. The Corps adhered to its levees-only policy, believing that upriver dams for storage held little potential, and indicating no apparent consideration of earlier recommendations to regulate development as a way of preventing flood damage.[23]

High water, however, continued to overtop levees, and in 1927 the Mississippi peaked at heights never before known, spreading across lowlands from Illinois to the Gulf of Mexico. Water submerged 16 million acres up to 30 feet deep and swelled to 60-mile widths. For perspective on this inland sea, consider that when we stand at the seashore, the horizon-line—so apparently distant—lies only 3 miles away. An estimated 250–500 people were killed and 500,000 relegated to refugee camps while flooding persisted through summer. Hardest hit were Mississippi's Black residents and other rural poor who—at gunpoint in some cases—were conscripted for sandbagging and related futile efforts.[24]

Damages were so profound that southern congressmen pressed for federal action in spite of long-held beliefs, dating to founding father Thomas Jefferson, that the role of the federal government in American life should be minimized. The South had long supported states' rights with a belief that sixty-one years prior had reached the whole way to the Civil War and its killing of 620,000 American soldiers back when our population was 31 million instead of today's 332 million. Few would even suggest that federalism was anything but repugnant to those of the White ruling class of the lower Mississippi basin. But there they were, in a blatant pattern of doublespeak destined to be repeated, pushing for Congress to pay to fix local flooding problems while demanding that the federal government leave the states alone in virtually every other respect.

Counter to their deep-seated if superficial political beliefs in autonomy, southern politicians after the 1927 deluge sought alliance with northerners to nationalize the flood issue in a way that nothing but war (with another nation) and a "natural" disaster could do. Congress in 1928 complied by broadly outlining the path for federal engagement with floods, and it directed the Corps of Engineers—no surprise—to increase work on levees.

Setting the stage for vastly expanding the federal role, spring of 1936 delivered heavy rain on top of deep snow in mountains upstream of Pittsburgh. Though the first flood recorded there by European settlers was in 1762, and although 115 notable high-water events followed, runoff on March 18 pushed river levels to record heights inundating Pittsburgh's classy downtown department stores to their gilded ceilings. The regional economy abruptly crashed while scenes of suffering lodged deeply into people's sensibilities locally and beyond.

The flooding of Pittsburgh and other cities in 1936 catalyzed Congress into passing legislation authorizing widespread federal dam building and the continuing construction of levees. Photo credit: Archives Service Center, Wikimedia Commons.

During this time, and against the dismal economy of the Great Depression, a battle of public policy ensued between Congress, the Corps, and Franklin D. Roosevelt. More thoughtful in regard to flood control than the average congressman, members of FDR's Natural Resources Committee in 1937 weighed options for dams and levees, which led them to think more broadly, and the members concluded, "It might be cheaper to prevent construction of new buildings in flood zones, and to gradually eliminate those now threatened as they become obsolete."[25] But like Ellet's recommendation of 1851 and the Corps' own advice in 1913 to relocate development away from danger, the Committee's observation represented the road not taken.

A debate hung in the balance between FDR's studious multipurpose approach to resource problems versus politicians' need to quickly deliver results that were figuratively and literally concrete.[26] Now realizing that levees would never contain the high water of cities such as Pittsburgh, built squarely on floodplains, and driven by political motivation from both northern and southern states, Congress passed the Flood Control Act of 1936—the first legislation establishing that flood control nationwide "is a proper activity of the Federal Government."[27] The law provided new direction including dam construction as a primary means of combating floods. Following the path of appropriations, a reluctant Army Corps quickly abandoned its levees-only approach and reinvented itself as the builder of dams. From its conflicted genesis, this mission grew into one of the more remarkable ventures, if not institutional obsessions, that one might document among federal agencies.[28]

Feeding the compulsion to grant the Corps more power and more money, even greater flooding occurred in 1937 when the Ohio River raged for its full 981 miles. A "1,000-year flood" buried Louisville 15 feet deeper than previous records.[29]

Though the path could have been different, the resolve to build our way out of flood problems dominated Depression-era public policy with powerful effects that linger. FDR had been reluctant to yield to the narrow structural approach, but the vertical gray walls of dams ironically came to signify "physical advertisements for the New Deal."[30]

With the 1936 Flood Control Act plus companion legislation in 1938, a busy and ambitious Army Corps produced study upon study documenting flood hazards along virtually every river in America. The agency inventoried thousands of dam sites and recommended construction wherever remotely feasible, Maine to California, prescribing comprehensive damming especially for the Appalachians, South, and Midwest where political support hardened in congressional currency known as "pork barrel." This meant bringing federal tax dollars (the pork) home in a feeding frenzy of government construction projects. Recognizing the greater dynamic at play, geographer and historian Rutherford H. Platt wrote that from 1936 onward Congress "would transfer much of the financial costs of disasters from individuals and communities to the nation as a whole."[31]

Major floods had again devastated Johnstown in 1936 and, with no letup, in 1937. Pop culture commemorated "flood city, USA" in one of the first Mighty Mouse animations, which portrayed that early superhero halting the rupture of city levees with his bare three-fingered cartoon hands. No less wishful, but grounded in civil engineering of the day, such as it was, a Corps project in 1943 channelized the Conemaugh River and reinforced its levees, leading residents to boast of Johnstown as the "Flood-Free City" that would suffer high water never again. A dependable axiom regarding flooding should, by then, have been apparent: never say never.

SERIAL FLOODING

Hundreds of dams and flood-control projects were built after 1936, but major floods continued to cause egregious damage, somewhere, virtually every year, affecting even normally dry regions. In 1938, for example, the usually desiccated Los Angeles River rose in foaming brown boils to kill dozens of people in famously sunny Southern California. The losses spurred plans to channelize the ephemeral stream in a cement straitjacket with hopes of flushing local runoff seaward and exporting it as fast as possible, ironically accomplished in the same era that the city built aqueducts hundreds of miles long to import every drop of water it could find from distant sources, even where other people had regarded the flow in their own regions as essential to their lives and livelihoods.[32]

Many floods later, and on the other side of the continent, Hurricane Agnes in 1972 stalled in a static low-pressure downpour dropping up to 19 inches of rain over mid-Atlantic states. Rivers peaked at historic levels across one of the broadest areas ever flooded in America, Roanoke to Buffalo, Wheeling to New York City. Agnes became the nation's costliest flood up to that time, with losses unmatched for a decade. In Pennsylvania alone, it destroyed 68,000 houses and left 350,000 people homeless.[33] Significant in the broader scheme of our response to floods, the Agnes losses pushed floodplain management, flood insurance, and other nonstructural approaches to the center of public policy (see chapter 7).

Likewise triggering policy reform, yet another Johnstown flood thundered through the Conemaugh valley when 12 inches of rain on July 19, 1977, caused not just one more dam above the beleaguered city to fail, but six, together killing 87 victims, destroying 400 homes, and crippling another 6,000. The losses pushed Congress toward long-overdue requirements for dam safety.

Floods of 1972 triggered new initiatives in floodplain management, zoning, and insurance. The Hurricane Agnes Flood has washed away most of a mobile-home park along the Susquehanna River's West Branch, and high water is receding, in this photo taken after the flood crest.

In the nation's crescendo of flood losses, the Mississippi in 1993 redefined the meaning of damage when it swelled in the "Great Flood" of American history. This covered 20 million acres, compared to 13 million in 1927.[34] Making such a disaster possible, if not inevitable, a winter's worth of snow at Rocky Mountain headwaters was followed by two months of rain saturating the soil, Montana to Louisiana. The river peaked at an astonishing million cubic feet per second, pushing lowland residents to board up windows, stack sandbags, and ultimately flee while rising crests inundated 100,000 houses and inflicted $15 billion in losses.[35] President Clinton declared disasters in 525 counties, including three repeat-declarations at one place: the Missouri and Mississippi confluence.

Dozens of dams had been built in the Mississippi basin at a cost of $25 billion, but virtually all of them filled with runoff prior to the 1993 crest. Unable to bear the months-long assault, 1,100 of 1,576 major levees failed, leaving thousands homeless. Unknown numbers of minor levees also broke and disintegrated. The flood ravaged 9 million acres of cropland, not so much by water, which eventually receded, but by smothering depths of sand that the flood deposited and left to stay. The federal government released $6 billion for relief, but the damage totaled much more. Scott Faber of American Rivers wrote that the flood was unique in forcing "the nation to question assumptions about roles and responsibilities for flood-loss reduction."[36] To do this the Clinton administration formed an Interagency Floodplain Management Review Committee assigned with recommending new federal policies (see chapter 7).

The West Coast had its turn in 1986 with record-high flows of the Sacramento and its tributary, the American River, peaking in what the California Department of Water Resources called the state's "greatest storm of record," curiously without much regard for the 1862 flood. The new deluge forced 50,000 people from homes while the governor proclaimed emergencies across thirty-nine counties. Then in 1997 rain dissolved deep snowpacks of the Sierra Nevada and again pushed the American River to levee-tops in the capital city. Just northward, levee failures near Marysville forced the largest evacuation in California history up to that time and destroyed or damaged 16,000 homes, all leading the state to adopt new initiatives to combat flooding.[37]

ESCALATING HURRICANES

Among the seventeen most-damaging hurricanes throughout US history, fifteen occurred since the year 2000. Katrina, in 2005, topped the charts, forcing evacuation of a million people when water submerged four-fifths of New Orleans. The flood breached levees where neighborhoods lay an ear-popping 20 feet below sea level, killed 1,200 people, and cost $169 billion in 2019 dollars.[38] Hurricanes reached northward as well, with Sandy costing $74 billion in 2012 and flooding far inland from the damage center of New York City.

In 2017 Hurricane Harvey amassed so much thermal energy from the Gulf of Mexico—warmed to a shocking 7.2 degrees F above normal— that it spawned American history's heaviest rainstorm. For five days the downpour stalled over southeast Texas; 52 inches of rain submerged one-third of Houston, including neighborhoods never expecting a flood to come anywhere near. Water rose 6 inches per hour, forcing 32,000 residents from homes and flooding the city's industrial corridor of 500 chemical plants and 10 refineries where benzene and a dozen other carcinogens escaped into streams, ditches, and neighborhoods.[39]

The 2000s became a "water world" of year-after-year blockbusters in flooding. In 2018 Hurricane Florence dumped 40 inches of rain that overflowed the Lumber and Cape Fear Rivers, leaving Wilmington, North Carolina, isolated. People died in flooded homes, from fires, by electrocution under toppled power lines, inside stalled cars, and beneath uprooted trees that crushed mobile homes. High water flushed through the state's hellish hog factory manure lagoons, poisoning water supplies, ruining properties, killing fish, broadcasting disease.[40] As if muddy water saturating homes and businesses were not enough, reoccurring floods highlighted the perils of storing unsafe and toxic materials on floodplains. But nothing quite compared to the risks created not just along, but *in*, our rivers. There, centered bank-to-bank, a plethora of aging, poorly maintained dams loomed as the greatest flooding danger of all.

FLOODS CAUSED BY DAMS

The safety of dams has concerned people ever since 1889 when the unprecedented carnage caused by the dam collapse at Johnstown shocked the nation. But beyond the private owners like those who had flagrantly failed to maintain that infamous structure, America's most respected and generously funded dam-building agencies have also been challenged by dam failures and near-catastrophes of colossal magnitudes.

To be clear, most of the dams that have collapsed and caused devastating floods were not built to provide flood control and were not built by public agencies. But some were. And all dam failures heighten the irony that this brand of disaster cannot be blamed even implausibly on an "act of God," as curiously claimed for so many floods, but rather on structures obviously built by people. Further, the damages and fatalities caused by failed dams are often greater than those of any other flood in a given area.

In 1928 St. Francis Dam, intended to serve semiarid Los Angeles with drinking water and completed with ribbon-cutting fanfare only two years prior, cracked like a broken egg, and its floodwaters killed 600 downstream residents overnight. This ranked second only to the 1906 San Francisco earthquake for fatalities by disasters in disaster-ridden California. But, while natural forces as inevitable as continental drift caused the notorious earthquake, faulty engineering and construction were clearly to blame for the St. Francis debacle, enshrined as the twentieth century's worst engineering blunder.[41] With an inexcusable blind-spot for errors, LA water czar William Mulholland descended in free-fall from status as a civic hero to public ignominy after the collapse.

Skipping any number of dam failures and fast-forwarding to 1972, Buffalo Creek Dam of Pittston Coal Company burst and caused two lower dams to fail, together releasing a flood of black coal slurry, like cold wet tar, horrifically killing 125 and leaving 4,000 West Virginians homeless, a preventable outrage illustrating the hazards of private and poorly maintained dams paired with inadequate regulation of both dams and the coal industry.[42] A parade of failures continued to the May 2020 collapse of two dams on Michigan's Tittabawassee River, forcing evacuation of thousands. An independent review panel found that those failures had been "foreseeable and preventable."[43]

The practical among us might conclude that we must maintain the dams we've built, just like we do with our infrastructure of roads, skyscrapers, and homes. But that's easier said than done. According to the Army Corps' *National Inventory of Dams*, 91,000 sizable dams (25 feet high holding at least 15 acre-feet, or over 6 feet high holding 50 acre-feet) have been constructed in the US, and many of the owners cannot now even be found. Yet all those structures age toward and beyond obsolescence while flood levels grow greater. Climatologist Noah Diffenbaugh at Stanford University explained that "the combination of aging infrastructure, older design guidelines, and an increasing probability of extreme events from global warming" was increasing risk of dam failures, stating, "More dams will collapse as aging infrastructure can't keep up with climate change."[44]

Hazards extend from private structures to major public dams, some justified partly for flood control. In 1976 the federal Bureau of Reclamation attempted to fill the reservoir behind Teton Dam in Idaho, newly completed, but the accumulating water on the volcanically porous landscape immediately penetrated cavities beneath the earthen structure, causing it to implode with dust, then explode with mud in a torrent killing fourteen people and scouring 100,000 acres of farmland that the impoundment was intended to irrigate. Awaiting below, where it spanned the much larger Snake River, American Falls Dam had been built in the early days of the Bureau, 1926. The need to upgrade the antiquated structure had been considered urgent, yet reconstruction was delayed. Subjected to the turbid onslaught from the emulsified Teton Dam, the larger dam downstream, containing a 23-mile-long reservoir, could have failed, triggering a domino-style collapse at any number of twenty additional dams awaiting in sequence for 1,000 miles farther down the Snake River, followed by another four mega-dams blocking the Columbia River down to tide line. Though a nail-biter to all, including Bureau of Reclamation managers, the antiquated American Falls structure held. Its rebuilding was accelerated and completed two years later.

Before Teton Dam construction had started, economists criticized it for an expected return of fifty cents on each tax dollar. To pad the flagging justification in the final days of the big-dam-building era, the Bureau credited 17 percent of the tenuous benefits to flood control, only to have the dam's failure cause more flood damage than an entire suite of federal dams in the Snake River basin had provided in flood protection over a span of decades. Taxpayers ended up covering $300 million in damage claims for a project that cost $100 million to build. Yet total losses topped $2 billion, all of which could have been avoided.

Dam builders at the Bureau had been warned about Teton's geological hazards by David Schleicher and others with the US Geological Survey—a sister agency whose expertise, one might think, would be welcomed. The geologist spelled out the disastrous scenario of piling dirt on top of porous substrata and expecting it to contain the weight and infiltration of a 17-mile-long reservoir. Failing to get much response, Schleicher suggested, "Since such a flood could be anticipated, we might consider a series of strategically placed motion-picture cameras to document the process."[45] The Bureau passed on that idea, but others rushed to the site in time to film the collapse, bequeathing us a video seen by millions and still being watched.[46]

Near-misses in dam security have reached startling proportions where failures would have altered entire populations and redrawn whole maps of settlement. In high runoff of 1983, Glen Canyon Dam's 710-foot rise above the Colorado River proved inadequate to contain springtime runoff given the operation plan being followed, and the integrity of the structure at one juncture depended on maintenance workers bolting plywood hastily bought at local lumberyards to add 4 feet on top of the elegant but deceptively inadequate dam. By mere inches this on-the-spot, seat-of-the-pants upgrade was the only thing that prevented water from overtopping the dam, which could have spelled ruin and a rearrangement of downstream geography through two states and two countries, a crisis masterfully described by Kevin Fedarko in *The Emerald Mile*.

How close did we come to losing Glen Canyon in 1983? John Keys, later promoted to commissioner of the Bureau of Reclamation, responded, "We came a hell of a lot closer than many people know."[47] The prematurely ailing Glen Canyon Dam has since been repaired and retrofitted to accommodate 220,000 cubic feet of water per second. An improvement, but without breathing room, this is less than the recently calculated 246,000 cfs expectation of a 500-year flood for the river upstream at Moab, not including the sizable inflow of the San Juan River, below.[48]

Concerns for safety lead directly to dam spillways—necessary features designed to accommodate floodwaters when reservoirs fill. Off to the side of the primary dam structures and slightly lower, spillways are concrete surfaces that allow rising reservoir water to escape downstream without overtopping the primary face of the dam and threatening catastrophic failure. Spillways of adequate design must be properly built and maintained in working order.

But at California's Oroville Dam, on February 12, 2017, they weren't, and 180,000 people downstream were ordered to evacuate—the largest evacuation ever in the US, under fear of dam failure. Another half million people farther downstream in Sacramento remained at home only because the Yolo Bypass—an overflow channel circumventing the river's main route through the city—had been acquired by the Army Corps in the 1930s in order to divert floodwater.

The Oroville drama played out while a crowd of state and federal officials, including top experts in averting dam failures, eventually knew they could be witnessing the greatest man-made flooding calamity of all time. Cracked concrete of the main spillway, eroding in massive chunks and propelled like cannonballs downstream, forced operators to switch overflow to an unfortified, secondary spillway consisting of a thin cap of concrete separate from the main spillway. But that soon resulted in an alarming rate of erosion gouging the bare hillside and ominously backstepping toward the reservoir's surface and its inland sea perched 771 feet above the valley below. Emergency efforts included the last-resort of helicopters dropping sandbags and boulders onto the riddled spillways—a low-elevation bombing exercise, hit-and-miss at best. Sheriff's deputies sprinted door-to-door ordering people to grab their pets and drive away, "Now."

Snowmelt and rainwater continued to pour into the brim-full reservoir. Corresponding torrents jetted out, unavoidably gouging the disintegrating spillways further with each hour of flow. Boulders and slabs of concrete continued to be spit downstream in the flood's crushing onslaught. Oroville approached the brink of a catastrophe dwarfing all others in a state synonymous with "natural" disasters. Except that this one was profoundly man-made and, as at Teton Dam, the problem had been explicitly predicted.

First as a polite inquiry in 2001, then as a pointed request, then an official intervention during the state-owned dam's federal relicensing process, a small nonprofit, Friends of the River, warned about Oroville's reliance on a bare hillside as a backup spillway, clearly stating that it "does not meet FERC's *Engineering Guidelines* for service or auxiliary spillways."[49] Not knowing where money for retrofitting would come from, and with apparent presumptions about the weather, flows, and safety, the California Department of Water Resources argued that the upgrades—requiring tens if not hundreds of millions of dollars—were unnecessary. Federal Energy Regulatory Commission officials agreed. But a postincident forensic team reported that the Department had neglected to act on reports of poor foundations at both spillways.

Though a close call, Oroville held. Reservoir levels began to recede while the forecast called for more flooding, prompting an all-out siege of repairs, ultimately costing $1.1 billion spread over two years.[50] At roughly one-tenth that sum, the original dam's construction had been contracted for $123 million in 1962.[51] The federal government ended up paying for

In 2017 California's Oroville reservoir filled, leading to uncontrolled dam releases that inflicted extreme damage to the structure's spillways. With fears of a flood induced by dam failure, nearly 200,000 people were evacuated downstream. Photo credit: California Department of Water Resources, William Croyle.

much of this add-on cost of the state's keystone water project, the entire episode reflecting *Grist* journalist Tim Kovach's incisive observation that "today's disasters are really just yesterday's unaddressed vulnerabilities."[52] Or, as hydrologist Nicholas Pinter of the University of California reflected, "with flood hazards, it's never the fastball that hits you. It's the curveball that comes from a direction you don't anticipate. Oroville was one of those."[53]

California, it turns out, can boast the best record among all states for addressing dam safety, meaning, principally, that all others are worse. Perhaps the most important conclusion to the Oroville crisis was stated in an independent analyst's postmortem: "The fact that this incident happened to the owner of the tallest dam in the United States, under regulation of a federal agency, with repeated evaluation by reputable outside consultants, in a state with a leading dam safety regulatory program, is a wake-up call for everyone involved in dam safety."[54] But did anyone wake up to the wake-up call?

Look no farther than Trinity Dam, just 140 miles north of Oroville. Lacking any spillway, whatsoever, Trinity was built by the federal Bureau of Reclamation to a height of 538 feet—certainly enough to pose dangers in the event of overfilling. When asked if spillways for such large dams are essential, Friends of the River's Ron Stork reflected on the Oroville disaster—which, by the way, he had warned about—and suggested that adequate spillway safety features are omitted "only on dams where the owner and regulators have the hubris to think they won't be needed."[55]

Another California eye-opener for risk, the Bureau's Auburn Dam on the American River was $200 million deep in construction when a nearby earthquake, ironically at Oroville, halted work, which was still limited to excavation for the dam's foundation, at that point looking a lot like a West Virginia mountaintop removal but without the coal to show for it. Unlike Oroville Dam's comparatively stable artificial Everest of rocky earthfill, a completed Auburn was engineered to be an elegant, double-curvature, thin arch of concrete, sculpturally shaped like the side of an antique teacup, its gleaming white intended to soar 680 feet into California's indigo sky. While construction was postponed for seismic analysis and redesign to a squatter, safer pile of rock and fill, stratospheric rises in cost sidetracked the Bureau's plans and, seeing no valid justification for what had become a multibillion-dollar gamble, the State Water Resources Control Board in 2008 revoked the Bureau's water right to it, pounding a stout nail into the dam's coffin but not quite burying the corpse. People including Ron Stork continue to try to do that, though they face lingering opposition from local boosters and congressional representatives who are unwilling to leave America's big-dam-building age behind, even forty years after President Jimmy Carter and Interior Secretary Cecil Andrus confirmed that it had ended.[56]

Having faced perennial proposals to resurrect Auburn, and confronting plans to build yet more dams on the San Joaquin River to the south, Eric Wesselman of Friends of the River urged California authorities to sensibly "prioritize funding" for the essential safety of existing structures instead of "wasting billions of dollars on new dams that are probably not going to get built anyway."[57] Above all a political realist, Ron Stork commented, "We still have canyons, and we still make concrete, so the controversies over Auburn and other dam sites are not over."[58]

Furthermore, look only to Oroville, again, for troubling conflicts regarding trade-offs in public spending. Owing to global warming and the storms it spawns, the Federal Energy Regulatory Commission this time ruled that the newly reconstructed spillway cannot carry the expected increase in runoff. The state's plans were considered "not acceptable," and a deadline was issued for resubmitting a new plan for federal approval.[59]

These dam failures and close calls do not describe isolated events. The nationwide picture regarding safety of dams shows 1,645 failures causing 3,600 fatalities between 1848 and 2017, averaging 10 failures annually, though the rate escalated to a troublesome 24 per year after 1980 owing to aging infrastructure, which, of course, gets older every day.[60] Indicative of the path we're on, and its fateful direction, three of the four most deadly floods of the 1970s were caused by dam failures.[61]

Unlike on highways, where drivers hit potholes and complain until the maintenance truck arrives with its cold-batch of asphalt, the deterioration, aging, and pothole equivalents of dams lie mostly out of sight, out of mind, and out of budget.

Contrary to the construction of all those dams, which had brazenly defied budget ceilings of both political parties for decades and survived consistent overruns at extremely high percentages, dam safety has been chronically underfunded if not starved of even a nominal allowance. When the National Dam Safety Program Act of 1996 finally passed, Pennsylvania had two inspectors for 3,000 dams and no money for repairs in the inevitable need that the inspectors found a single problem. Alabama had no safety program, at all, but 2,273 dams.[62] In 2018 the Army Corps' *National Inventory of Dams* rated 2,300 sizable dams as "high hazard" (failure would endanger life and property) or as "unsatisfactory" or "poor" in condition, meaning vulnerable to damage or collapse. Many of the private owners could not be found, leaving safety concerns up to government agencies, taxpayers, and the grim budget realities of no budget at all.

In an analysis that engineers likely trust, the American Society of Civil Engineers in 2021 graded the safety of American dams with an unflattering "D." Among eighteen types of infrastructure evaluated by that organization of infrastructure builders, it was dams, levees, stormwater drainage, and roads that scored lowest.[63] The Society estimated that costs to fix all dams needing repair would run $65 billion, but no sources had anywhere near that kind of money. Likewise, needed levee repairs would cost $50 billion—an implausible investment among levee districts, some of which may be strained to buy sandbags needed during regular crises.[64] Putting these numbers in perspective, Eric Halpin, retired chief of dam and levee safety for the Corps, reflected that the agency had been investing only $500 million a year in infrastructure repairs at dams. "Everyone knows it's more cost-effective to prevent damage than to repair it, but it's not in the national will yet."[65] Just what that requires is a good question. In another analysis, Jeffrey Mount of the Public Policy Institute of California projected that upgrades to only California's dams and levees would cost $34 billion—nowhere to be found, even after Oroville's flirt with disaster.[66]

The Biden administration's much-championed but compromised and narrowly passed infrastructure bill allocated $585 million for dam repairs in 2022.[67] The federal funding was widely welcomed, though it represents yet another subsidy for private and public entities that built dams under the premise that the projects would be economic bonanzas—or at least pay their way—and with the promise that they'd shower their communities with enviable bounty—or at least be safe. But deadbeat dams get a free ride to the future when their exorbitant costs will come due to people who will never have received a penny in benefits.

Other than a lack of funds for repair or removal, and in addition to the failure of dam owners to take responsibility for their structures, one of the key reasons for increasing risks of failed dams—and of flood damage generally—is called "hazard creep." Going to the heart of this book's core topic, it's not actually the hazard that creeps, but rather new development that gets cobbled up directly in harm's way. More and more people build homes where they'll be flooded if and when dams fail. Thus, the danger grows even if the dams don't deteriorate—which they do—and even if the inflow from storms isn't projected to increase—which it is. The lack of a growing maintenance budget might well define not just the tired theme of denial, but a crisis that's out of control and destined to come home to roost with unprecedented disasters unthinkable today.

Following the Oroville malfunction, and with reflection on shortcomings, California officials got the message and required dam owners to assess potential failures, to map the extent of downstream inundation possible, and to draft emergency action plans. In contrast, typical Federal Emergency Management Agency floodplain maps—the regulatory standard nationwide—lack even elementary information on flood risks posed by dams upstream of homes and cities.[68]

The ultimate irony is that in some cases, unsafe dams now pose flooding dangers shockingly greater than the dangers earlier posed by flooding rivers before those dams were ever built. At tremendous cost, we've made flood prospects worse.

A purely economic imperative—even without concern for compelling ethical and practical issues of safety—often means that the best approach to unsafe dams is to get rid of them. Thus, as the twentieth century closed and a new one began, America had moved beyond the era of dam construction and into one of dam removal, with many sites to choose from. Nearly 2,000 dams were removed as of 2022 (remember, America has 90,000 sizable dams still extant). Most were privately owned and small, but not all. Glines Canyon Dam, rising 210 feet and privately built for hydropower on Washington's Elwha River, was demolished in 2014 to bring back salmon. Thousands of unsafe dams remain, with impending failures posing great risks of floods. The structures still stand, though usually without purpose, all posing burdens on jurisdictions that own them or that have inherited the costs, not to mention the risks to people who live below.[69]

The organization American Rivers champions the removal of unsafe and purposeless dams and announced 57 removals accomplished with that organization's help in 2021 alone. Though immediate safety was not the primary concern, four large hydropower dams endanger salmon in the Klamath River of California but will be dismantled in history's largest river restoration project, scheduled for 2024.

INCREASING STORMS AND THE MISSISSIPPI'S FOREVER FLOOD

American history's largest siege of flooding, so far, occurred in 2019. People had thought that the Mississippi's "Great Flood" of 1993 would never be exceeded, at least not in a single lifetime. But 2008 brought the second "500-year" flood to the heartland in fifteen years. In 2011 levels crested even higher in some places. Yet the worst always seems yet-to-come, and likely is.

After a wet 2018 and record-breaking ice-buildup in January 2019, deep snowfall blanketed the Mississippi River's vast headwaters, followed in March by what the TV weather people have grown fond of calling a "bomb cyclone" of wind and rain. Runoff jammed waterways with ice, pushed flood flows over banks, broke dams, cleaved levees, and spilled uncontrolled across what had been considered flood-proofed land spanning parts of eleven states.[70] In the end, 2019 set records as the wettest among 124 years of monitoring and as the greatest long-term siege of high water in American history.

Most unusual about the 2019 floods was that they persisted winter-to-fall with scarcely any letup. Record rainfall just kept coming. Overflows stretched across lowlands for miles from the banks of the Mississippi, Missouri, Arkansas, and other rivers. Forget the typical bell-curve of increasing river levels, cresting sharply for a day or two of crisis before coasting back down to normal. The 2019 hydrograph looked more like the back of a whale, and the acreage that farmers could not plant, all summer, doubled previous maximums. Newscasts revealed heroism and tragedy. Just one of these dual dramas involved a Nebraska farmer driving his tractor to rescue stranded motorists when a bridge beneath him collapsed.

Summer 2019 served as the most urgent wake-up call yet regarding increasing floods, not only to people exposed to danger, but also to financial institutions that seemed to have sent their actuarial staff home without pay and blatantly bought into the public's ambient level of denial about future floods. In what sounded like dredgings from archives of the Great Depression, the Federal Reserve Bank of Chicago—a heartland institution—announced in autumn 2019 that 70 percent of its farm clients were unable or acutely challenged to pay back their loans. The year's losses and the projections for future flooding suddenly sent ripples through the financial world that makes debt-driven modern agribusiness possible.[71] Those ripples expanded as the 2019 deluge left economic footprints far larger than high water itself. For example, across whole midwestern states, foul flood seepage polluted a million private wells, each an economic loss, but more poignantly a personal tragedy to a family.[72]

From spring through autumn in 2019, the Mississippi River repeatedly rose to record levels and flooded in durations never before anticipated. Saturated farmland here, upstream from St. Louis, typified much of the river basin's floodplains.

Disruptions extended beyond millions of acres of corn and soybeans, beyond hundreds of towns and cities, and into the critical realm of national defense. One-third of the Offutt Air Force Base in Nebraska was flooded, with 137 structures inundated, some 8 feet deep. Not just a National Guard base for optional weekend trainings of Guard enlistees, this nerve center of the US nuclear arsenal is key to Department of Defense Strategic Command. The hazards had been known, dating to a 2011 flood and earlier, but many vulnerable facilities had not been relocated.[73]

Navigated since before Mark Twain's brief but storied tenure as a riverboat pilot, the Mississippi in 2019 swelled too high, too fast, and too dangerous for anyone to negotiate. Far from the realm of picturesque paddleboats dating to the old bard's *Life on the Mississippi,* one modern barge, 35 by 200 feet, carries the equivalent coal or grain of seventy semitrucks, and one towboat pushes not 1 but 42 barges, collectively a serious problem if out of control in swift current with bridge pilings and other barges to dodge. So, at its second-highest level in history, the river at St. Louis remained closed to commercial navigation for months. This meant no wheat or corn shipping out and no fertilizer or fuel shipping in—effectively a whole economy on ice.

Worse, levees failed at hundreds of sites that had been blithely regarded as protected.[74] By the end of June, eleven midwestern states—every one of them with politicians disinclined to talk about their chronic dependence on the federal government—sought federal disaster funds. Lower Mississippi crests exceeded those of the 1927 flood's infamy in spite of billions in taxpayer money that had been spent to lower flood levels throughout a ninety-two-year campaign aiming to avoid exactly what was starkly happening in 2019.

With exhaustion unseen after lesser floods, victims across the Midwest began searching for a way out by moving. "We can't keep this up and make a living," said Michael Peters, a corn and soybean farmer along the Missouri River.[75]

Flooding events in 2019 also spanned the nation, from the Russian River in coastal California where navigating downtown Guerneville required oars, to low-country of the Southeast when Hurricane Dorian dumped 15 inches of rain and forced evacuation of 360,000 residents from the Cape Fear River floodplain. With no letup, the increasing power and number of hurricanes through 2020 illustrated the ominous force of global warming.[76]

Worldwide, flooding problems grew even worse. Floods and storms displaced 21 million people yearly on average, 2008–18—three times the number of victims displaced in that interval by the combined wars and political conflicts that dominate nightly news.[77] In India and Bangladesh a hurricane in May 2020 forced authorities to prepare for evacuation of 5 million people. Two times the population of Chicago suddenly had to abandon their homes. From May to July 2021, unimagined flash floods crippled cities in Germany and ripped through the Philippines, India, London, and Detroit, each leaving a legacy of destruction no one ever wanted to see again. Yet we did.

In 2021 Nashville was buried in 7 inches of rain, its highest two-day accumulation ever. Other towns in Tennessee got a staggering 17 inches in twenty-four hours. Then, as a dark reminder, Hurricane Ida flooded parts of New Orleans precisely on the sixteenth anniversary of Katrina. After costly investments following that disaster of 2005, the city's heavily reinforced levees held, though flooding devastated other communities and the storm wreaked damage up the East Coast to New York, where 3.2 inches of rain fell in one hour, shattering by 1.3 inches the previous record, set only eleven days before. Subways became Rivers-of-Styx and forty-six people died, many in the comfort-turned-horror of their own basement apartments in our nation's largest city.[78]

Even with the photogenic hype of hurricane winds bending palm trees in half and the impressively dusty rubble of earthquakes, floods accounted for 75 percent of presidentially declared disasters in 2021.[79] Ground-zero among these, the Mississippi basin has endured flood risks since the prehistoric mounds of Cahokia, but where the onslaught used to occur once per decade, it might now occur once or twice a year. Direct losses nationwide increased from $4 billion annually in the 1980s to $25 billion, both in 2020 dollars, all in spite of many other billions of taxpayer dollars spent to control floods.[80] Our approach was clearly not working.

In 2022 Hurricane Ian inflicted the most damage ever caused in storm-ridden Florida with losses of $60 billion from winds, waves, and river flooding. It didn't have to be so bad. Under earlier political leadership, in 1985 the state had wisely established a program to manage risky development, and it was making notable headway when gubernatorial candidate Rick Scott in 2000 derisively called the risk-reduction program a "jobs killer," blazing the trail for a Republican legislature to flatly abolish the responsible agency in 2011.[81] The real jobs killer was Florida's vulnerability to flooding, witnessed when Hurricane Ian in 2022 rendered tens of thousands of workers jobless.[82]

In January 2023 California floods caused evacuation of 30,000 people and disaster declarations in thirty-one counties. Then in March, unimagined 12-foot depths of fresh snow in the mountains above Los Angeles awaited the dreaded rain-on-snow event, which came.

The floods of the 2000s were turning the Earth into a very different place from anything America and the world had ever experienced. Even though we had suffered through the requisite lessons of flooding again and again, little evidence suggested that we had learned what we needed to know.

LESSONS FROM THE HISTORY OF FLOODING

Memories of historical flood disasters imprint indelibly on all who suffer through the cycles of high water. People who have endured floods remember rowboats navigating Main Street, rescuers floating to second-story windows, mud smeared on and impregnating everything, everywhere. After the storm has past, many flood victims find that their problems have only begun, and that recovery remains a long, excruciating, debilitating, often impossible process.[83] Once they've seen it, smelled it, and felt it, many victims and even distant witnesses have concluded that our vulnerability to floods is a problem that must be solved, or at least dealt with responsibly.

In efforts to do that, we built levees in the nineteenth and twentieth centuries, and upping our game after 1936, we built 400 costly flood-control dams. The Hurricane Agnes Flood of 1972 catalyzed efforts for floodplain management and an improved National Flood Insurance program. In 1977 the second great Johnstown flood spurred Congress to address dam safety. In 2019, 2020, 2022, and 2023—may as well say every year now—the hurricanes and floods of the twenty-first century bring awareness that the changing climate will unleash a floodwater apocalypse of repeating onslaught. Resolve among some people has coalesced around the need to do something other than what we've been doing for two centuries—to do something more effective, more affordable, more possible, whatever that is.

Anyone who considers this history of flooding and who notices its trends and trajectories might ask, in an era of wetter climate and rising floods, what will the future bring? How should we respond today? Can floods be controlled? Can vulnerabilities be reduced?

As I pursue answers to those questions, I'm reminded of a hopeful scenario expressed in the writings of environmental historian Ann Vileisis: "Informed by history, we can remember the trade-offs already made and turn away from the mistakes and misunderstandings of a time when we knew no better."[84] Let's hope for that.

To give our ancestors the benefit of the doubt, flooding's history might be regarded as one that describes times when—excusably or not—many people knew no better. It's also a tale of nature and culture, a recap of willful ignorance along with enlightening science and hard-nosed economics, a recounting of efforts to reform hopeless schemes that cling to myths of the past and to denials of the present. It's a story unfortunately imbued with the triumph of shortsightedness and of private enterprise outcompeting community welfare.

Missing—totally missing in this documentation of fact and myth—is the message that the workings of nature, and the natural laws upon which we most-fundamentally rely, all require not that we eliminate floods, but that we recognize their continuation as inevitable and also as essential in the timeless cycles of the Earth. In the next chapter let's step back from the losses and turn to those cycles of creation and of life.

3 Rivers Need Floods and Nature Needs Floodplains

THE BEAUTY OF FLOODS

As the rain increased, the flow of the river intensified with resonant, muscular force, heard but not quite seen through a curtain of conifers and other woodland wonders along the McKenzie River as it foamed down western slopes of Oregon's Cascade mountains, a wild place in the tamest of times, and wild at a viscerally primal level when the river rises. The flood was on.

With nothing in the way, with no homes or lives at risk, it was a natural phenomenon as inevitable as the turning of the planet and as essential as the cycling of water from sea, to sky, to land, and back to sea.

In late March, daylight hours equaled those of night, and as the curtain of darkness dropped, clouds thickened even more. Needing a lot of water, western red cedars, black cottonwoods, and white alders all thrived along the river and across its valley. The ongoing hourly transformation of that forest into a wetland was palpable as the river pulsed into wider spaces of the fertile plain formed by floods in the deep past. The ongoing storm, the flush, and the drenching power of it were all unleashed in a fundamental act of creation that was, at once, alluring, frightening, and beautiful.

The river rose by the hour and even by the minute. The sound of the flow had grown beyond any friendly swish or bubble and into the realm of roar, as if an underground, or maybe ethereal voice were making itself known to reclaim ancient status. "I'm still here," the river seemed to be saying in a bigger-than-life version of what's normal. Then the mysteries of a damp dusk turned into a dripping darkness hiding everything. Soon the terrestrial would become the aquatic, shifting a critical balance in the way the world was presenting itself, reminding me that the roots of life, including distant ancestors of my own, came not from solid ground but from water in the sea. The overflow of the McKenzie indicated how dominant water on Earth really is, not just on 70 percent of the globe that is the ocean of our origins, but even across our landscapes. We, and floods, together, go way back.

I spent the night nearby, parked on higher ground, sleeping in the dry comfort of my van, which I've outfitted for extended expeditions. I didn't really care much if I got isolated, temporarily, by water spilling over low spots in the highway upstream or downstream of my safe spot. Intimately close to the elements, yet effectively sheltered, I listened to rainfall wax and wane on the van's metal roof, a cadence underlain by resonant rhythms of a river rising in crescendo nearby.

Returning to the McKenzie at daybreak, I discovered a different world. The rain had stopped. High water had peaked under cover of darkness and in its wake left a renewed landscape. The moss had brightened to vivid green, the ground had softened like a wet sponge. The flow that had steadily thickened in a cocktail of muddiness the day before was now shifting back toward green as the concentration of creamy silt began to dilute. Fresh sand had settled in a previously dry back-channel, which had grown into a lively stream of its own and persisted, now, even as flood levels slowly receded.

Making my way down along water's edge to where the back-channel reunited with the main stem of the river, I saw the two flows join in a turbulent eddy. I knew that beavers thrived in the wetland world nourished by floods because delectable sticks that had been peeled and partly eaten spun in the circular current there. Hefty logs, buoyed on the flood's crest, had crunched against other fallen trees and become stranded in a logjam. One fat timber had lodged across the slough, creating a walkable bridge to the island, though I decided not to risk the slippery surface over the McKenzie's still-turbocharged secondary channel.

The flow in the night had delivered fresh gravel, now mounded in glistening piles at the river's edges and underwater. Each deposit—like shining coins of natural capital generously gifted to the bottom of the stream—would form essential habitat for cutthroat trout, and also for steelhead and salmon if they could make their way upstream after multiyear sojourns in the ocean where, feeding on greater resources of the marine environment, they had grown larger than their fluvial life alone would have allowed.

Nature's high water had supplied the river's backwaters, saturated the porous soils of the floodplain, recharged groundwater underfoot, and delivered mineral nourishment in the form of alluvial sand and gravel. The combined scene illustrated one of Earth's great natural cycles that had been continuing ever since the river valley wasn't even a distinct valley, but rather unarticulated, harsh, youthful terrain dating to explosive volcanic eruptions that millennia ago had formed the spacious geography all around.

A path lured me farther downstream where windows through the forest and its vertical columns of cedar trees repeatedly opened to the rushing river. Like a garden hose that in the backyard one might aim directly into loose soil and in that way scour out a microdepression, the flood force of the river had, at a vastly greater scale, scoured or deepened apparently depthless pools within the water's channel. Likewise, riffles and rapids had grown more turbulent because newly deposited cobbles or freshly tumbled boulders now added more friction to the flow in shallow places, bleaching the river's surface foamy-white. As most floods do in upper reaches of mountain watersheds, this one had renewed both pools and rapids in a rhythm of alternating stairsteps: gentle, then steep gradient. Pools, then riffles, rapids, and pools again.

Not entirely constructive with its deposits of sand, gravel, and cobbles, the power of the water had also carved into earlier layers of sediment with centrifugal force pushing against the outsides of bends, lopping off hungry bites of soil and forest duff and washing them downstream for redeposit on sandbars at the insides of bends below, all in balance and direct complement to the erosion from above.

Scenes like this appeared along the McKenzie everyplace lacking houses, roads, pavement, farmed fields, levees, riprap, dams, and other artifacts of what we do along rivers. I knew that in towns far downstream the flood might have brought misfortune and misery, but up there where the water flushed through the woods, it brought gifts and life to the world around it by refreshing all things natural along the shores, all things native through the emerald canyons and valleys, and all things living, including trees, forests, fish, and birds that depended on the flow of water.

SEEING FLOODS DIFFERENTLY

Ever since people began telling creation stories of the past, we've celebrated themes of bending a hostile world to serve our desires and needs. Adam and Eve cultivated their Garden of Eden, complicated and compromised as it was with snakes. Skipping ahead quite a ways from that story, we have the Army Corps of Engineers in Johnstown, Pennsylvania, standing up to its own serpent, so to speak, by "flood proofing" what had been called "Flood City, USA." However, another story, different from the river rearrangements of the Corps, and far more fundamental to the workings of the world, was waiting to be heard, and in many circles it still is.

That story contrasts sharply with the mainstream vision of malevolent storms and savage floodwaters monopolizing myth, history, the news, public policy, popular metaphors, and people's deepest emotions. Those popular historic accounts, reiterated after floods, speak to tragedy stemming from "acts of God." But another story tells us that floods are not only inevitable, no matter what we do, but also beneficial and, in fact, essential to life on Earth.

If that notion were even imagined—which it wasn't through most of the past century and longer—it would have been anathema where I grew up along the Ohio River, listening to an older generation recount the 1936 flood upstream in Pittsburgh, and likewise at the Susquehanna where I personally encountered the Hurricane Agnes deluge in 1972.

Being trained in landscape architecture and working as a land-use planner in the 1970s, I knew that our approaches to floods had not worked as well as we needed them to do. But the idea that floods were actually good was ground-breaking and, to use a phrase of the time, mind-blowing.

Hydrologists, ecologists, and biologists had been working along this line for a while, but I remember the day and the hour when the value of floods came into focus for me, when the light bulb beamed, so to speak. Attending a conference focused on rivers in 1975, I sat down in a darkened church basement in Washington, DC, to watch a film about floods. Like other people there, I expected some retelling of dangers, tragedy, and heroism when rivers overtop their banks, all illustrated with graphic images of suffering and loss. But I emerged fifty minutes later with a completely different view. The film articulated a new and informed way of understanding floods and opened a fresh outlook in the minds of many who were there.

The versatile director, writer, and cinematographer Lincoln P. Brower taught ecology at Amherst College, and he introduced the film and answered questions about it afterward. Sporting wire-rimmed spectacles and wearing baggy pants and a sweater fit for the chill of his home in New England, the understated but congenial professor might have met anyone's expectations of a university scientist. His considerable renown as an entomologist stemmed from unrelated research on monarch butterflies, which migrate via unimaginable instinct to and from Mexico on journeys that span not just one generation of the insects' lives, but several, and in each, the individual butterflies know where to go—sight unseen—to fulfill their fair share of the migration route. In Brower's presence, one would more likely think science "nerd" than environmental "activist" or "water policy reformer."

However, several years before, Brower had enlisted with a group having the mundane title of Connecticut River Basin Coordinating Committee, chaired by the Army Corps of Engineers and assigned to address floods of the Connecticut River. This largest of New England's freshwater arteries happened to curve within view from the Brower home. Responding to the Corps' nine-volume, 25-pound report advocating 7 major dams and 196 smaller ones, all for the purpose of "stopping" floods, Brower and other biologists from nearby schools in the college-rich region of Massachusetts took note. They intuitively knew that no one could build 200 dams and not affect the ecosystem. So they organized, as academics do, an open-minded symposium, in this case for the practical purpose of reviewing the

Corps' plan. What the professors learned alarmed them, and the public's blind-spot about flooding alarmed them even more.

Ahead of his time, and perhaps stemming from his awareness of the orange-and-black monarch's spectacular beauty when fluttering by the tens of thousands in migration, Brower also recognized the value of visual media—that people had to *see* the beauty of nature and also the problems that natural systems face. With his initiative, the Connecticut River cadre of New England professors secured Rockefeller Foundation funding to produce a film, *The Flooding River: A Study of Riverine Ecology*.[1] Among its opening lines, Brower proclaimed that floods have become "dangerously over-controlled."

"*What* did he say?" may have been the initial response of many, under their breath, as it was of myself while the film rolled onward. In the documentary the professor proceeded to explain that "the widely held assumption that flooding is ecologically harmful is incorrect." Quite to the contrary, Brower emphasized, "flooding is a beneficial and renewing natural event. It's responsible for producing and maintaining the biodiversity of river ecosystems."[2] I knew about floods, having narrowly evaded high water and, in rubber boots, shoveled a lot of mud myself. But here was Professor Brower telling me that we *need* floods. I mustered the patience to listen.

He built the convincing case that floods are essential for a healthy shaping of the riverbed, of its shores, and of its entire corridor, and that floods provide organic nutrients critical to complex networks of fish and wildlife. Overflow areas are essential to migrating birds in springtime. The biologist noted that 80 percent of our oceans' commercial and noncommercial fisheries depend on estuaries where fresh water and salt water mix in a wealth of nutrients that are delivered, as it turns out, mainly by floods. The movie won awards and caused me, at a young age, to think more deeply about floods beyond the usual images of mayhem, wreckage, and loss.

Taking up the same torch that Lincoln Brower had held decades before, and with more data to back him up, biologist John McShane wrote that "floodplains offer more value to society per acre than any other landform."[3] Their relatively small area accounts for "25 percent of all terrestrial ecosystem service benefits"—the basic processes on which our essential air, water, and living resources rely.[4]

Newscasts are typically loaded with pejorative language blaming floods for misfortunes. The reality of those painful descriptions owes not to the fundamental nature of floods, and not to their effects on the natural world, but to the vulnerabilities created by people who find themselves—for whatever reasons—in the path of rising waters.

FORCES OF CREATION

To go along with Lincoln Brower and say that floods are good and beneficial is a bit of a stretch for many of us, so let's back up to one of the basic ways that the world works: the hydrologic cycle. We've heard this before: water evaporates over the oceans, prevailing winds push the clouds up and over land, they cool and lose their ability to hold water vapor, and the moisture condenses. Rain and melting snow then gather into rivers flowing back to sea, where the process repeats timelessly. What we don't usually hear about in this climatic rerun is the fact that the inevitable variability of water returning to sea includes floods causing rivers to overflow, as well as droughts causing streams to trickle. However, in the long-run, and in whatever terms nature dictates, random events have averaged out in ways that for eons have nourished life.

In simplest terms, a flood is what happens when more runoff occurs than the channel of a river can accommodate. This has occurred countless times along every river. In fact, the rules of hydrology and morphology—meaning the science of nature's structures including the shapes of waterways—produce river channels that are *made* to flood. There are really no other options given Earth's landforms and weather. The relative depth of channel, height of banks, and width of overflow space are all determined by uncompromising physics of flow stemming from a river's combined and interacting geologic substrate, climate, and runoff. The result yields rivers that on average overtop their banks every 1.5 years in uplands and a bit more frequently in coastal zones.[5]

For perspective on an even larger picture, consider this bedrock truth: floods have been a principal force in shaping the earth and making it usable to people. Let's back up again, simplify just a bit, and imagine nature's most fundamental sculpting of the land as a two-step process. First, the earth has been pushed up into hills and mountains, and second, the surface of that earth has been worn down and contoured into what we now see.

The built-up part of this process owes to geologic forces of plate tectonics, seismic seizures, and erupting volcanism that have lifted the Earth's crust into seven continents, thousands of mountain ranges, and millions of hills. The worn-down part of the process is governed mostly by forces of water—mainly rivers that carve those upraised features into the landscapes we now know. This earth-shaping process is, first of all, one of subtraction. Rivers create valleys and canyons by removing particles of soil and rock from mountains and whole continents, much as Michelangelo removed flakes of raw marble to reveal the exquisite figures he carved within. Flowing water accomplishes this in a seemingly random but actually orderly fashion by picking up silt, sand, and rocks and carrying or rolling them downstream with the flow.

But that's only half of it. The earth-shaping process of floods is also one of addition. Floodwaters freight the suspended soil and rock downstream only until the river's current slows down enough to drop its muddy weight of rocks, then gravel, then sand, and finally, where the current comes to a halt, fine-grained silt. These deposits grow to form the nearly flat floodplains alongside or just beyond the banks of rivers where the water, when high, spreads out and pauses.

Over the course of geologic time, these flood settlings account for virtually all the bottomland in valleys and also the extended flats of coastal plains near seashores, plus the gradual tilt of the Great Plains, no less, as they cant imperceptibly downward from Rocky Mountains to Mississippi valley—that's the way floodwaters have rearranged rock and soil after eroding it from above.[6]

Almost all this land-shaping is done not during normal rainstorms or the small-to-medium flows, which are low-energy events with little effect on erosion and deposition, but rather during action-packed floods. In the Los Angeles River, an extreme example, the bulk of sediment transport between 1938 and 1969 occurred during only a few days. Robust power is required to move great amounts of soil and rock as earthy, agitated slurry. In most rivers, only floods have this force, even though the high water may last only a day or two per year, or just once per decade or even less in drylands. On the other hand, exceptional floods in large rivers such as the Mississippi can continue nonstop through entire rainy seasons with enough push to carry weighty loads of suspended soil seaward the entire time.

So, through both erosion and deposition, floods have shaped the landscapes we now know, use, and require as living space. In this sense, floods are not the battleground of destruction, as portrayed in conventional history, but rather among the most essential and omnipresent forces of construction on the globe. If you want to see the surface of the earth being shaped or, for that matter, created, look no further than the rise of floods. They've made the Grand Canyon, Los Angeles Basin, and eastern coastal plain, to name just a few classic American landscapes, plus virtually every flat bench and terrace perched above riverbanks, which is where much of our best farmland and many town sites lie—some still in the path of floods, but many sitting well above the rivers' modern crests and on floodplains that had formed in ancient times when the riverbeds were higher than they are now.

While these long-term notions about the importance of flooding offer little solace to anyone with property troubled by high water, it's worth noting that we wouldn't even *have* that property if it weren't for floods that created it.

Flood*plains* are defined as the land next to a river or hydraulically connected to it.[7] The Federal Emergency Management Agency cites 7 percent of the United States as floodplain or, specifically, as "floodprone."[8] For data here, our usual geographic authority, the US Geological Survey, defers to FEMA.[9] The floodplains' relatively small acreage is—from environmental, ecological, economic, and cultural perspectives—among the most important parts of the landscape, and what we *do* with that acreage involves some of the most critical land-use decisions we collectively make, for better or worse, as we'll see in the chapters that follow.

THE SHAPELY FIGURE OF A RIVER

Being the great force of creation that they are, floods shape whole landscapes, including the critical layout of a riverbed and its shores—a subgeography whose condition we seldom think about but that we might regard with the same relative importance we attach to the health of our own fluid-circulating arteries.

Taking note of the physics involved, ecologists regard rivers together with their floodplains as "disturbance" ecosystems, which means that they thrive on and actually require periodic upheavals, even though these may not appear beneficial or productive at the time. We readily see evidence of disturbance in high water's erosion at the outside of a river's bend and its deposition on the inside. Even though the edges may seem a bit rough, the flooding process renews the shape and condition of the shores, diversifies habitat, and nourishes forests.[10]

In the channel itself, the formation and maintenance of riffles, rapids, and deep pools all depend on flood flows, as we saw along Oregon's McKenzie River. Those riffles and rapids are essential to streamlife because they aerate the water with bubbles that boost oxygen supplies, which are critical to the "breathing" of fish and other creatures. This is important but critically so in waters burdened by algae stemming from overloads of sewage and farm waste. The overabundant algae in badly affected streams ultimately dies and rots, consuming oxygen in the process, which leads to anaerobic conditions favoring foul bacteria and noxious species ranging from carp that displace native fish to red flatworms that grossly multiply in low-oxygen waters. By aerating the water, rapids counteract these plagues.

In essence, fast bubbling flows purify rivers, which might seem intuitive to many of us who feel an intrinsic sense of refreshment and revitalization, if not joy and excitement, while sitting at the edge of a rapid or waterfall. Likewise, and at the opposite end of the hydrologic phenomena of creation, the pools that are scoured out by flood flows are essential to streamlife because they offer deep-water shelter for fish evading predators coming from land or sky. The pools are resting places where currents eddy out, and they offer protection that might be sensed by anyone relaxing at the edge of a quiet riverfront scene. Vitally important in the age of spiking heat waves, deep water in a river's pool acts like the insulation of a cooler shielding a picnic lunch from direct sunlight. This respite from increasing temperature grows more crucial for native life, including trout, steelhead, whitefish, and an entire food chain of cold-water-dependent species. Where lacking the deep-pool chill, brook trout in the East and cutthroat trout in the West suffer when water temperatures top 65 F, and those famously feisty fish go limp and perish at 75 degrees if thermal refuge cannot be found in pools or at cold tributaries.

Streams that are dammed or diverted often lack the flood flows needed to create and maintain the valuable pool-and-riffle sequence, and so their gradients even out as the pools gradually fill with silt and as the rapids gradually disappear with slow but incremental erosion, all leading to a constant, biologically boring glide of current rather than the enlivening pool-and-riffle stairstep. Consider, too, that without flood flows, silt is not washed downstream to be deposited in beneficial ways on the insides of bends, at sandbars, and across deltas. Instead, the silt builds up directly in streambeds, smothering clean gravels needed by fish for spawning. That buildup also clogs space in the channel that's needed to accommodate high flows, causing greater flood levels when big runoff eventually occurs. Lacking requisite high flows, rivers that have been dammed for decades become congested with silt and also entangling brush and debris that's no longer flushed away.

Floods are so fundamental to the morphology and life-history of rivers that the concept of a stream as a water-filled channel is not quite right. A better definition of *stream* or *river* includes not just the path where water flows, but also where it overflows—the floodplain.

NATURE NEEDS FLOODS AND FLOODPLAINS

Floodplains are the most biologically valuable of all terrestrial habitats.[11] Typically made of coarse sandy soils and grainy silt with porous spaces that become filled with water, floodplains act likes sponges when rivers overflow. This moisture feeds plants rooted there and also sinks deeper to form cold reservoirs of groundwater that sometimes extend back from the river for miles. The subterranean water serves not just plants and wildlife, but cities, towns, and rural homes; half our population depends on groundwater for domestic and municipal supplies tapped by wells. In the Merced River basin of California, 46,500 acre-feet of water per year could be made available by allowing flood flows to recharge groundwater basins.[12] That's enough for 90,000 households in a state severely stressed by water shortages. Groundwater provided by floods later resurfaces as

Floodplains including riparian forests and wetlands such as these in northern Minnesota rank as the most critical of all landforms nationwide for fish, wildlife, groundwater infiltration, water quality protection, and natural overflow basins during storms.

springs, which augment creeks and rivers otherwise dropping low in summer.

Many floodplains include marshes and swamps. These wetlands enhance wildlife and water quality. Wetlands excel in absorbing floodwaters, thus lowering peak flows in the rivers, all with beneficial spinoffs economically. Researchers found a savings of $745 per year in avoided flood damage with every acre of wetland protected. Savings grew to $3,240 in developed areas.[13]

When water spreads over floodplains, bacterial action breaks down nutrients and allows them to settle on land rather than polluting streams. On the other hand, where woodlands and floodplain open spaces are developed or plowed down to the riverbanks, water is degraded by nutrient surpluses coming from overfertilization of fields, feedlot runoff, and erosive farming practices.[14] Along the Rio Grande, for example, floodplain forests filter out two-thirds of the nitrates in overfertilized runoff coming from farms, which can otherwise cause serious problems ranging from fish-kills to blue-baby syndrome.[15]

Beyond critical issues of water supply and ecosystems, floodplains rank among our most important recreational assets. People choose riverfronts as places to walk, run, camp, bike, or simply relax.

FLOODS AND FORESTS

Forests that thrive on floodplains rank among our most valuable habitats for wildlife and for ecosystem maintenance including defense against a heating climate. Important, for example, are eight species of cottonwoods and sycamores that grow to be our largest deciduous trees. All these grow best—if not exclusively—on floodplains.

Biologists call cottonwoods a keystone species—many other plants and animals depend on them. The cottonwoods, in turn, depend on floods. For germination, these impressive trees typically need either a freshly scoured sandbar or a new silt deposit. To take advantage of those, the trees' seed dispersal is evolutionarily programmed to occur just after springtime flooding. Opportunistic seedlings quickly send taproots deep to draw nourishment all summer from what biologists call the "hyporheic" zone where water seeps into sand and gravel alongside streams. In this way the trees provide a riparian corridor of fruitful wildlife habitat, not just in moist regions but also across plains, steppes, and deserts that account for 40 percent of America. Forests there are mostly limited to the thin green ribbons of floodplains winding through comparatively barren terrain. Unfortunately those forests are dwindling owing to diversions of water and to dams that prevent flood flows from renourishing the riparian landscape.

If not dependent on floodplains, many species of trees do best in deep alluvial soil deposited by high water, ranging from the great sycamores and silver maples of Appalachian hollows, to the Southeast's bald cypress swamps, to the Olympic Peninsula of the Northwest where the largest western red cedars anchor roots along rivers. As the tallest trees with the greatest density of biomass on Earth—far greater than tropical rainforests—redwoods reach skyscraping heights on floodplains of California's

Even during high runoff up to the tree-line, the upper McKenzie River in Oregon flows with clear water from its undeveloped, forested headwaters. Naturally occurring logjams—created by floods—provide vital habitat for native fish.

Smith River, Eel River, and Redwood Creek. The revered trees tap moisture that accumulates in root zones during floods.

The floods' gifts of old-growth forests are essential not only to fish and wildlife, and not only to many of us who cherish strolls-of-a-lifetime through cathedral groves of towering trees, but also to our recovery from global warming. Carbon dioxide, released into the atmosphere by our burning of fossil fuels, causes overheating of our climate. Reversing that process, photosynthesis in trees consumes that same atmospheric carbon and beneficially converts it back into solid carbon in the form of trunks, limbs, and roots. Forests of floodplains—and by association the floods themselves—are among our best allies in combatting global warming. Other vegetation and sediment in related bogs and riparian wetlands likewise sequester carbon in great amounts.

Trees along rivers and streams eventually die, and some fall into the water and are swept away by floods to downstream bottlenecks where the driftwood accumulates in logjams. These are not just piles of woody debris that congest channels and snag fishing lines, but vital contributors to ecosystems. The logjams back up water to form pools where gravel settles as fishes' spawning beds. In-stream logs also redirect currents to create riffles, and they provide anchorage and food for invertebrates eaten by fish, which also hide under logs to evade predators including wading bears and diving osprey.[16] Eventually high flows push some of the logs downstream to estuaries where wood-boring organisms digest and recycle the fallen trees, enriching marine food webs far offshore.[17]

With all this in mind, biologists tell us that fallen trees must be reinstated into waterways to restore streams and the life that depends on them. But trees large enough to lodge themselves in the currents have mostly been cut and trucked to sawmills. The benefit of logjams is one among many compelling reasons to spare and reinstate mature riparian forests and to safeguard floodplains from both logging and development.

Streams need logs, logs come from big trees, big trees grow on floodplains, and both the trees and the floodplains on which they grow require floods in order to exist.

WILDLIFE NEED FLOODS

Not only charismatic species such as fish-eating eagles and bears, but entire food chains depend on floods. High waters in overflow channels and wetlands collect organic detritus as nursery grounds for phytoplankton—microscopic bacteria, algae, and plants that account for half the world's photosynthetic activity and thereby half the globe's essential oxygen production. Most phytoplankton live in oceans, but their presence in rivers underpins critical food webs linked to insects and then to larger creatures the whole way up the food chain. Nourishing that ladder of life, many native plants depend on what biologists call the "flood-pulse advantage," which benefits native species and also purges invasive exotic plants and animals from rivers and their overflow zones.

Only 1.5 percent of America's original floodplain habitat remains in healthy condition, yet biologists credit riparian or riverfront land with supporting 35 percent of endangered species and the richest diversity of life-forms.[18] In the semiarid Southwest, 65 percent of all species rely on riparian areas whose health depends on periodic flooding.[19] Freshwater habitat of streams and their immediate shorelines—covering only 2 percent of the Earth's surface—have the highest diversity of animal species. Diving deeper, biologists report that phylogenetic diversity— a measure of organisms' genetic relationships to each other and thereby a key to species' long-term health and evolution—is twice as high for freshwater and related terrestrial habitat than for any other part of the Earth. Thus, preserving streams and their floodplains can protect more species than can protection of any comparable area on land or at sea.[20]

Floodplain forests that filter out silt and pollution make the life cycles of many fish species possible. Fish native to North America evolved with flood flows and manage to evade or cope with the flush of fast currents when they occur. High water also provides cues for spawning and for aeration of redds—the gravel "nests" where fish eggs of many species incubate and hatch. Many spawning areas lie off the streams' main channels, on floodplains. In the Mississippi River, for example, fifty fish species rely on seasonal overflow habitats for spawning. There and elsewhere fish need periodic high flows that deliver food including insects and other nutrients that fall into the water from inundated trees and shrubs. Overflow channels and connected wetlands provide further refuge where fingerlings can evade predators such as alien largemouth bass that lurk in main channels. Anyone who likes fish has got to like floods.

Riparian zones also rank among the richest habitat for birds. Floodplains of the Sacramento and Mississippi Rivers are principal reasons that two of the great migratory flyways of North America exist.[21] Where floodwaters flow, waterfowl dive for fish or dabble for aquatic plants at the surface, herons spear minnows and reptiles at the edge and nest in cottonwoods, and wood ducks shelter in cavities of riverfront sycamores. Songbirds thrive in the canopy and in willow thickets that require the high-water pulse.

Mammals living at river's edge depend on riparian habitat and the floods that nourish it. Beavers maintain entire riverfront wetlands by building organic dams with their sculptural art-form of sticks, grass, and mud. Their hydraulic management of small streams benefits fish and other wildlife, and also arrests downcutting of currents that can otherwise incise stream channels and reduce them to sterile sluiceways.[22]

The Environmental Protection Agency reported that 75 percent of western wildlife need riparian habitat, and the figure for the East may not be much different. Yet 70 percent of that habitat has been lost to development or badly altered.[23] Recognizing habitat needs, hydrologist Philip Williams noted, "There is almost a perfect overlap between the measures needed for implementing an effective flood management strategy and those needed for meaningful restoration of fish and wildlife."[24]

OCEANIC CONNECTIONS

Floods provide critical silt and nutrients to estuaries at the edges of the oceans. This became evident at the mouth of Washington's Elwha River following elimination of two large dams in 2011–14. The salmon's return to the newly dam-free river succeeded as the main goal of the removal, realized immediately, but a pleasant surprise came with reclaimed wetlands and delta deposits that the restored floods delivered where the Elwha entered the Salish Sea. The river's high-water flush now serves food webs expanding far offshore. Similar floods at virtually all estuaries provide nutrients, minerals, structure, and biomass to the timeless circle of life at the fertile fresh- and saltwater interface.[25]

Sea-level wetlands of flood-deposited silt also buffer impacts of battering ocean surfs during storms and provide the front-line of defense for property lying landward, even during hurricanes. However, dams have robbed many rivers of their natural silt flows that historically arrived via floods. Likewise, flooding rivers deliver the sand that makes up our beaches, provided flood flows continue. In California, the Ventura River's Matilija Dam, 200 feet tall, has starved ocean beaches by trapping seaward-bound sand. At Los Angeles' doorstep, Rindge Dam, blocking Malibu Creek a century ago, filled with sand in twenty-five years, depriving Southern California beaches while Pacific currents swept existing beach deposits away, leading to shoreline erosion and oceanfront houses collapsing into sea.[26] If you like beaches, you have to like the floods that provide those beaches with their sand.

FLOODPLAINS STORE FLOODWATER

Beyond the habitat, ecological, and geographical values they provide, floodplains most fundamentally serve as a place where floodwaters can go. They are, in effect, natural reservoirs, working for free. But they get no credit as such on the ledgers used to calculate benefits of flood-control efforts or to compare flood-reduction potential of floodplains versus dams. As the Corps stated in 2022, "There are no comparative studies between valley [floodplain] storage and reservoir storage."[27]

However, when high flows spread out across a valley and into adjacent wetlands, they decrease the level of flooding downstream. They also knock down damaging spikes of the hydrograph, more safely spreading the duration of high water out over time.[28] In all cases, floodplains, by definition, accommodate floodwaters, and in some cases spacious floodplains along miles of river can naturally accommodate and store enormous amounts of water with benefits that remain uncounted, all without the downsides of dams.

INEVITABLE AND BENEFICIAL

That floods are beneficial to the functioning of rivers, riparian land, and whole geographies beyond them was simply not on anybody's mind throughout most of American history. But now it is. Ever since Lincoln Brower's film and onward to the contemporary work of ecologists, hydrologists, and economists, a new message resounds that flooding isn't inherently wrong or sinister, but rather an indispensable part of how nature works. To solve the problems of flooding and reap intrinsic rewards from our rivers, this crucial natural process needs to be left intact and restored.[29]

Knowing the importance of natural floodplains, and that most floodplain management historically focused only on reduction of property losses, the Association of State Floodplain Managers in 2008 wrote, "We need to marshal unprecedented forces to preserve and improve the natural functionality of our floodplains and coastal areas and protect the resources they provide. In doing so, we will also mitigate damage and losses that floods bring to society." The Association called for us, first, to protect surviving floodplain open space from development; second, to "remove existing development from flood-prone and environmentally sensitive areas whenever possible"; and third, to "rehabilitate and restore degraded riparian and coastal resources."[30]

Though flood issues remain absent from general public consciousness and even from most environmental literature, activism, and campaigns, the loss of floodplain assets arguably ranks as the most important problem in the stewardship and care of rivers today. Unlike pollution, which can often be remedied in relatively short order by eliminating its source, the problems of development on floodplains become permanent. Once paved, riparian areas usually stay paved. Consider, too, that while we justifiably celebrate gains in conservation with removal of even one unneeded dam and the return of even a few miles of river to free-flowing status, we lose far more miles daily to new development infringing on floodplains. Those losses go unnoticed, in part because they occur so commonly.

Floods are not only inevitable but also essential to the great cycles of the Earth and its systems of life. At a gathering of Northern California scientists on the fiftieth anniversary of that region's epic 1964 flood, hydrologist Mike Furniss summed up this view with words that might be as unexpected by some people today as they were when Professor Brower created his landmark film: "Floods are something to celebrate."

Beauty, of course, might lie in the eyes of the beholder, and seeing what might be an unexpected beauty in floods depends on the beholder knowing the ways and the imperatives of the natural world. Being attuned to all of that raises the question, why would we choose to go against, rather than with, the inevitable forces of a hydrologic cycle that's so much greater, so much more timeless, and so much more powerful than we are?

Nature and all that depend on it need floods. Yet floods can damage whatever we build in their way. So let's next consider what we've tried to do to stop that damage from occurring.

4 Dams for Flood Control

THE PROMISE AND THE REALITY

STOPPING FLOODS, TAMING NATURE

Floods can be understood as the inevitable outcome of a perfectly natural process—the hydrologic cycle—and thereby a phenomenon of beauty, power, and creation underpinning nothing less than the way the world works. But not everybody sees it that way. And to stop floods, who can argue with the logic behind building a dam?

Here's how it goes: too much water for our liking comes at once, causing floods. If we contain the flow in reservoirs behind dams, we can reduce peak levels downstream, then later release the stored water harmlessly, or maybe even beneficially. This way we can control or eliminate floods. That's to say, we can control nature, which is what's causing the flood problem.

With that argument—case closed—the age of dam building for flood control began, as we saw in chapter 2 with the Flood Control Act of 1936. Many billions have been spent in pursuit of this goal. For decades few questions were asked in spite of uncertainties and a lot of stones left unturned. Whole cities developed with the promise of no more floods. Entire segments of the economy have depended on that expectation. People gravitated to an engineering-driven, congressionally authorized,

Army Corps of Engineers version of Moses's parting of the waters: below the dams the land will be rendered dry; above them will be permanently flooded by the reservoir pools.

Congressional authorizations for federal agencies to address floods in some other respects had come earlier, but after the northeastern flood of 1936, policy shifted from one that had favored levees as the nation's flood-stopping strategy, to building dams. Though explicitly opposed by the Army Corps for the whole previous century, dams were cobbled up after 1936 even on sites considered barely feasible, and all on a playing field sharply tilted in favor of a culture-wide war against high water. Congress opened the floodgates of funds to build dams in an era principally remembered for the Great Depression and its dire economic needs that, sandwiched between two World Wars, left little room for introspection. Unasked, unanswered, a list of questions accompanying unforeseen consequences of dams arose in the decades that followed.

Who pays? Who benefits? What degree of flood protection will be assured? What does "assured" even mean? What level of floods could be expected at the dam sites? What alternatives might be possible, practical, or economic? What were the ironic risks of flooding by failures of the very dams that were intended to eliminate the floods? What were the downsides and the collateral damage to society and the environment from building dams? How will the dams be fixed, rebuilt, or dismantled once they become unsafe or useless? Even with the Dust Bowl and the floods that followed it, people during the Great Depression did not expect to encounter a heating climate worldwide and its intensified storms, but how does that reality affect our operation of dams today?

A LEGACY OF DAMMING

Before 1936, and without federal help, a few cities had crafted their own flood-control plans involving dams. In 1908, following a flood where the Allegheny and Monongahela Rivers collided to form the Ohio River at Pittsburgh, a commission chaired by the ketchup magnate H. J. Heinz aimed to tame floods with reservoirs catching water upstream. But even the industrially booming Pittsburgh, with the likes of Andrew Carnegie's fortune and philanthropy, was not up for the capital investment required, and regarding federal intervention, Americans nationwide lacked the will to fund anything as provincial as flood control for a particular city.[1] Pittsburgh's high water was Pittsburgh's problem.[2]

People in the Miami River valley of Ohio took a different approach after a 1913 flood that motivated residents to grab their own bootstraps and pull on them to form the Miami Valley Flood Control Association. The group contracted engineer Arthur E. Morgan—later first director of the Tennessee Valley Authority—and collected $2 million from 23,000 private subscribers to build five dams and channelize the Miami through Dayton and Cincinnati.[3] Local responsibility remained the American mindset until the deluges of 1927 and 1936, which catalyzed a seismic shift in the federal presence regarding floods.

A curiously blended hybrid of military and civilian manpower, the Army Corps of Engineers had maintained channels for commercial navigation of the Mississippi for steam-powered paddleboats during the golden age of river travel in the 1800s. Broadening the purview, the Flood Control Act of 1917 added flood control and hydropower to Corps responsibilities. Agency leadership adhered to a long-held view that dams cost too much for uncertain benefits. Meanwhile levees offered clear assurance seen in floodwalls separating city streets from rising waters.[4]

However, with the collapse and widespread overtopping of Mississippi levees in 1927, and with painful inundation of eastern cities nine years later, political resistance to dams evaporated. Declaring that dam construction for flood control was a responsibility of the Corps, the 1936 Flood Control Act stated, "It is hereby recognized that destructive floods upon the rivers of the United States, upsetting orderly processes and causing loss of life and property . . . constitute a menace to national welfare . . . it is the sense of Congress that flood control on navigable waters or their tributaries is a proper activity of the Federal Government." Declaring war on the "menace" of floods unleashed what would become one of the most massive manipulations of nature and landscape in history.

The Corps' first major dam had been authorized in 1935 on West Virginia's Tygart River, not for flood control but—in keeping with the Corps' earlier marriage to commercial barging—to augment low flows troublesome to coal shippers on the Ohio River. This assignment, by the way, paved the path for later and greater subsidies gifted to fossil fuel industries and ultimately landing us where we are with global warming. But more germane to our story, just one year after assigning the Corps to the Tygart, Congress unleashed the agency on a spree of dam building to subsidize those who might benefit from flood-control structures. The Corps targeted Appalachia, with its convoluted topography, climatological recipe for rainstorms, and easily navigated politics, but no part of the country escaped the dam-building boom except the remote icebox of Alaska.[5]

Pushing the dam-building trend along, the growing prowess of professional engineering—glamorized by art-deco designs of sleek steel and concrete—was cleverly associated with a yearning for something "great" in the darkness of the Great Depression. Forces coalesced to make megaprojects compelling in the public eye, however unexamined the final results remained. Dams that blocked the flow of rivers became an uplifting act of progress, perceived as symbolic and real. Once the door to federal funding opened, it swung wide and stayed that way as long as money propped that door open by flowing through it in a stiff and unrelenting current of congressional appropriations, which no one even bothered much to question for several decades.

With hardships of unemployment outdone only by those of high water, nothing could have been more popular than the government spending millions to hire men, move earth, and bring nature to heel by building dams. The Army Corps was the savior, riding waves of respect and power, and with impunity it built dams not even approved by Congress. One unauthorized "project modification" relocated a dam 100 miles to another river and another state: from West Virginia's West Fork River to Pennsylvania's Youghiogheny.[6] With advanced technology that made huge dams possible, the 1936 flood-control legislation along with kindred hydroelectric dam authorizations ignited an era of big-dam construction that defined grandiose public works for five decades. Time after time, at ribbon-cutting ceremonies when floodgates were slammed shut and waters began to back up, society took a deep breath of pride, believing that a fearsome world had suddenly been tamed. The satisfaction became addictive. One dam led to another. Smaller ones led to larger ones. Virtually every elected official wanted one. Or more.

Shredded by the Depression, the nation's earlier beliefs in limited government fell by the ideological wayside with Franklin D. Roosevelt's presidential victory and his legendary "One Hundred Days." He created the Tennessee Valley Authority in order to erect dams, but also to strive for a less-celebrated and far-less-endowed regional goal of "comprehensive resource management." Eventually nine dams blocked the arterial Tennessee River, plus twenty-three were built on tributaries, mostly for flood control. TVA permanently flooded 635,400 acres including 10 percent of the Appalachians' finest farmland—numbers that might, in some other realm, have been called into question given that only 7 percent of America is flood prone, and given also that prime farmland is precious because everyone, after all, has to eat, while anyone, theoretically, can choose not to live on a floodplain.

One TVA dam forced removal of 3,500 families, many of them farmers, each sacrificed property a personal tragedy with permanent scars on the land and among the victims, though unnoticed in the national hype for damming. The resulting social relocation escaped the public outrage bound to occur with even a fraction of comparable eviction of people by government edict today. While mountain farmers lost their homes to reservoirs, one city, Chattanooga, soaked up 85 percent of the combined dams' flood-control benefits.[7]

To build dams in the West, Congress had already created the Bureau of Reclamation in 1902, and across sixteen dryland states (then Texas was added in 1906) a facade of rugged individualism and blustery antigovernment sentiment quietly surrendered to dependence on a federal agency. The Bureau funneled billions of dollars westward, principally gleaned from densely populated eastern states, and spent them on irrigation projects beyond the 100th meridian—the Great Plains' boundary of semi-aridity that had served as a red-flag of warning to settlers. A federal policy of enticing and subsidizing farmers to move to the West's drylands somehow escaped the wrath of eastern politicians whose own agricultural economies suffered as a result.[8] More important to our story here, the Bureau of Reclamation didn't squander opportunities to bolster justifications for its irrigation dams with added claims of flood-control benefits. The Bureau, for example, completed Shasta Dam and California's largest reservoir on the Sacramento River in 1945 with a promise to stop the flooding of farms and cities, including no less than the state capital.

DAMS FOR FLOOD CONTROL

Flood control dominated the Army Corps' budget through these go-go years until 1965, when navigation for barges on the Mississippi and a few other large rivers again topped the Corps' ledger owing to inflation of maintenance costs, aging of navigation dams, and inadequate locks for larger strings of barges that the shipping industry wanted in order to bolster profits and keep it ahead of its competitor: privately funded railroads.

Nearly every catastrophic flood from 1936 through the 1960s was followed by an expansion of federal dam building.[9] More floods meant doubling down with more dams, and prospects for construction seemed to be as unlimited as did the threats of high water. Ted Schad, who had been an influential Senate committee staff director and Bureau of the Budget officer, said, "If ever there was an age of dam building, the forties and fifties were it. In 2,000 years historians will look at these structures the way they look at the pyramids today, and they'll call us 'The dam builders.'"[10]

After failing to justify even one dam, Pittsburgh's flood-control ambitions gained traction and the city became the beneficiary of eleven Army Corps dams. A lynchpin, 184 feet high on the Youghiogheny River, was completed in 1948 with overwhelming support in the "Steel City," where high water of 1936 was to flooding what 9/11 became to terrorism. Unnoticed in the hype, the Appalachian town of Somerfield disappeared under Youghiogheny Reservoir, as did an idyllic farming valley settled 170 years prior by Revolutionary War veterans who had been given fertile frontier property in return for unpaid military service no less than a frigid winter at Valley Forge. In 1982 former Somerfield resident Elsie Spurgeon recalled with a vividness that fundamental memories instill, "Nobody wanted to move. We were very happy there. It was an old town. It meant a lot to us. We figured it would be that way forever."[11]

Largest of the projects to protect Pittsburgh, Kinzua Dam on the Allegheny River arose in 1959 but only after displacing the entire Seneca Indian Tribe, one among six Iroquois Nations whose combined government had served, in part, as a model for the American constitution, although the Indians' charter for governance notably did not include provisions for eminent domain.[12] Recognition of the Indians' 10,000 years of living along the Allegheny, and of what the United States government interpreted as "ownership," dated to the Pickering Treaty of 1794—oldest of all American treaties negotiated with Native Americans. Signed by Chief

Cornplanter and George Washington himself, the document recognized Seneca ownership of 30,189 acres and stated that the Indians' land "shall remain theirs, until they choose to sell the same to the people of the United States." Skeptical of this ultimatum to give away all the rest of northwestern Pennsylvania and southwestern New York and to keep only a relatively small parcel along the river, the Seneca first rejected the treaty. They eventually relented after Pennsylvania's Quakers stepped forward, as peacemakers with reasonable confidence, to guarantee the word of our nation's "father" and first President. With that truce, and with the option to relinquish their "ownership" and "sell" their remaining land being beyond the pale in the Indian psyche, the tribe endured along the Allegheny for another 142 years following George Washington's signature.

Thus, it should be no surprise that after Congress authorized the Army Corps to build Kinzua Dam in order to reduce flood levels in Pittsburgh, 200 miles downstream, the Native people objected. Sympathetic, fearless, and literally towering over his colleagues but unable to sway their votes, Pennsylvania Congressman John Saylor challenged Pittsburgh's dam proponents who occupied his neighboring district, saying, "Apparently you have become so calloused and so crass that the breaking of the oldest treaty in the United States is a matter of little concern to you."[13]

Rallying to back their ancestors' word from a century-and-a-half earlier, the Quakers arranged for a study of alternatives by hiring Dr. Arthur E. Morgan—the same engineer who had designed the Miami Valley Project and served as TVA's esteemed first chairman. After careful analysis Morgan proposed to divert Allegheny floodwater into a glacial depression in the adjacent Conewango basin, reporting that more water could be stored there at less cost than that of Kinzua Dam. Not unsympathetic to Pittsburgh's needs, the Seneca supported the plan in what we would now call a "win-win" scenario. For an "outside" review of Morgan's proposal, the Corps hired former Corps employees in a firm that regularly worked for profit under contract for the agency. They reported that the Seneca's plan was feasible but would cost $91 million more than the dam. To save this sum, however disputed, a district court ruled in favor of the Corps, and in 1959 the Supreme Court concurred.[14]

The Seneca were promptly moved to southern New York and their valley flooded. Pittsburgh has not suffered a major flood since, and the combined

78 DAMS FOR FLOOD CONTROL

eleven flood-control dams reduced crest levels there 12 feet during Hurricane Agnes in 1972, according to Corps reports. Thus far, Kinzua Dam has worked, but the alternative—even according to the Corps—would have worked just as well. Or better and at less cost according to Morgan. But in the golden age of dam building, alternatives virtually never displaced plans for Army Corps dams, widely considered the gold standard for flood control. *Alternative* was scarcely in the dam builders' lexicon, at all.

400 DAMS

In spite of a few glancing blows, dam building remained popular among mainstream Americans for five decades. The Corps' flood-control system for the Ohio River, for example, grew to 72 impoundments.

Nationwide, 400 large dams were constructed to stop floods (many thousands of others were built for other purposes and by others). The Corps also piled up 14,100 miles of levees and dug 11,000 miles of channels to flush water away faster in places where it could not erect dams to hold the water back and make it move slower. At its flood-control heyday, the agency claimed $3.50 in benefits for every tax dollar spent.[15]

Under contract by the agency, historian Joseph Arnold crowed, "These remarkable engineering projects today comprise one of the largest single additions to the nation's physical plant—rivaled only by the highway system."[16] Yet the inflation-adjusted average of flood damage consistently increased and topped $9.4 billion annually, 2011–20. The principal reason for increasing damage is that increasing numbers of people buy homes on floodplains because they believe the federal dams have made them safe.

FLOOD CONTROL FOR SPECULATIVE DEVELOPMENT

Speculative real estate benefits on floodplains—distinct from benefits of development already existing at the time the dams were built—were calculated increasingly by the Corps to justify federal dams. It's one thing, critics observed, to claim benefits for a dam that stops flood damage to houses and infrastructure existing downstream, such as Pittsburgh in

1959. But it's another thing to claim benefits from development that does not exist and that could be located elsewhere with no need for expensive public works. Yet, as the feasible dams were completed, this is how the Corps increasingly justified flood-control projects.

Remember, the Corps itself argued until 1936 that dam building didn't make economic sense. Dams could only be justified if they protected enough buildings from flooding, so the more buildings that could be counted, the more robust the justification. This rationale not only drove the movement to claim benefits for speculative building that had not yet occurred, but also encouraged more building of homes and businesses on floodplains after the dams went up.[17]

Look at the numbers. In the 1940s some 10 percent of the Corps' claimed flood-control benefits came from protecting, not existing, but rather prospective, future floodplain development, according to a 1966 Task Force on Federal Flood Control Policy. No one blew the whistle, and the Corps in the 1960s justified more than 40 percent of upcoming flood-control dams' benefits by assuming development that was only a gleam in real estate agents' eyes and that could have been avoided through zoning or precautionary choices by people seeking homes that lacked flood hazards. In 1965 the Bureau of the Budget calculated that 50 percent of the Corps' flood-control benefits claimed for new dams owed to speculative development.[18] In the 1968 Flood Control Act, the number rose to 85 percent. To put this upward march of numbers in context, the Corps justified many flood-control dams by slim margins of benefit-versus-cost—sometimes only a percentage point or two making the difference in building a dam or not. So inflating the benefits tipped the balance sheet from failure to success in gaining spending approvals by Congress.

An entire program justified by the goal of reducing the suffering of flood victims had, in effect, morphed into a real estate juggernaut benefitting not flood victims so much as speculators waiting in the wings for taxpayers to make low-value floodplain land valuable, if not a windfall to those on the inside-track of decisions. Land-use regulation of floodplains was anathema and would have killed the party, so to speak, under this compelling paradigm. Controls on floodplain development would, of course, negate the speculative profits to be made by real estate interests there, and land-use limitations would obviate the positive benefit-cost ratios that Congress required of the Corps.

Opponents of Gillham Dam on Arkansas' Cassatot River pointed out that 74 percent of the dam's "benefits" were for flood control even though significant development did not exist for 49 miles below the dam—far beyond reach of the dam's flood-control abilities. The hotly contested project was completed in 1975. At Applegate Dam in Oregon, scarcely any development existed downstream the whole way to the Pacific Ocean, and so 80 percent of flood-control benefits were credited to speculative development in riverbank forests and fields, home principally to muskrats, minks, beavers, and birds. Amid advertising that local economies depended on the project—a totally bogus claim—Applegate Dam narrowly survived axing by a local initiative and was built in 1980.[19] At that late date, the whistle was finally blown, and when questioned about speculative benefits in 2022, a Corps spokesperson reported that the agency "does not include open land or the conversion of open land in the planning process, *except* under certain circumstances" (italics included in the statement).[20]

Regardless of protocols and of how the rules of justification might have been stretched, shrunk, or shifted to facilitate new development, and with perceived protection from floods, dams and levees contributed to disasters of increasing magnitude because expectations of safety led more people to build in zones that remained vulnerable to floods. As the National Water Commission reported in 1973, regardless of $8 billion spent on dams and levees up to that time, annual losses increased "because flood plain usage was increasing."[21] Without critiquing dams or his agency, Corps historian Martin Reuss fifteen years later sensibly wrote, "It might be that the federal government, before constructing flood control projects, ought to require or at least urge local interests to pass ordinances that would regulate flood plain development in order to eliminate the possibility of a major disaster from a future flood."[22]

A GROWING CRITIQUE

The Corps' accounting leaves little doubt that flood levels downstream from some dam sites were reduced on many years, averting economic losses. But, as the Indians at Kinzua attest to this day, the gains were not without costs, many going uncounted. Those sacrifices began to be noticed increasingly through the 1960s and 70s.[23] People questioned the logic of trading temporary and occasional flooding of land downstream of a dam for permanent and full-time flooding of land upstream of the dams, and of unconditionally calling the result "flood prevention."

Scarcely acknowledged during the Corps' dam-building heyday, but increasingly noticed as time passed, the agency ignored most fish and wildlife qualities of rivers while inflating the projected fish and recreation values of reservoirs. This even occurred where ample flatwater recreation was already available behind existing dams, and even where plentiful, unique, and sometimes imperiled native fish were traded for exotic ones that were dependent not on rivers but on artificialities of flatwater behind the dams. The Fish and Wildlife Coordination Act of 1934 marked a step in the right direction, requiring dam builders to consult the US Fish and Wildlife Service, but recommendations remained nonbinding and easily ignored. Many dams also caused depleted and overheated flows downstream where water was diverted or where it was annually held back to fill reservoirs—an outcome largely ignored for decades but eventually considered noncompliant with the 1972 Clean Water Act.

More subtly but ubiquitously, impoundments trap organic matter otherwise on its way downriver, and thus starve lower river segments of carbon necessary for streamlife to thrive. Outflows from dams on the Missouri River, for example, lack vital nutrients and turn the downstream aquatic environment comparatively sterile. Wildlife agencies, in a largely futile effort, resorted to dumping brush into the river to compensate for diminished nitrogen and carbon. No such knowledge of chemistry was necessary to see and comprehend that salmon, migrating upstream to spawn in the Northwest and Northeast, were abruptly halted at the concrete faces of dams. These fish had been keystone species of celebrated importance ecologically, culturally, and economically. In predam years, salmon numbering in the many millions had returned from sea to reproduce in rivers and streams, and then to die and thus fertilize both the flowing waters and also adjacent shores when bears, eagles, and other predators consumed the fish and recycled them onto the landscape. In this way, salmon served as the free vehicle for return of nutrients from sea to land—a process supporting even the growth of towering forests along the Northwest's famously wooded waterfronts. The barriers of dams, and the flatwater they backed up on the Connecticut and then Columbia, Sacramento, and other rivers, however, reduced the world's greatest salmon migrations to token numbers. Many native runs went extinct, with others projected to join them.[24]

Other effects of the dams were similarly unexpected, undesirable, and mind-boggling in pervasive outcomes reaching far beyond any hopeful prevention of floods. For example, when high turbid flows of rivers stall in the dams' flatwater, silt settles to the bottom of the reservoirs, causing the operation and flood-absorbing capacity of those dams to be reduced as the accumulating silt—mud if you will—displaces reservoir volume intended for floodwater. Elephant Butte Reservoir on New Mexico's Rio Grande had become one-quarter filled with sediment by 1997, and 5 million more tons accumulated annually.[25]

Dams trap about half of the famously viscous loads of sediment in the Missouri River, legendarily "too thick to drink and too thin to plow" but now evolving with silt-settling flats actually thick enough to plow in upper ends of the reservoirs where the silt settles out and preempts space formerly available to floodwaters.

With ramifications going far beyond those dams becoming white elephants filled with brown mud, the silt they accumulate is needed downstream for balancing stream-bank erosion, sandbar building, maintenance of the pool-and-riffle sequence, and continued existence of lowlands as significant as Louisiana, where the Mississippi has historically been called America's greatest "land-making machine" for its role creating America's largest delta, wholly composed of the river's sediment deposits at and near sea level. The silt that's needed to maintain that acreage is now trapped and wasted in upstream flood-control reservoirs. As a result, salt water encroaches, and its landward advance is aggravated further by geologic subsidence of the Mississippi Delta, where communities, going back to the Acadians and further, are now literally sinking into an encroaching sea because silt once delivered by the river is no longer deposited in roughly equalized compensation for the sinking landmass.[26] Louisiana has lost 2,000 square miles of wetlands—larger than the state of Delaware. More disappears every year in a real-life Atlantis located not in fictional waters of the Atlantic Ocean but at today's Gulf of Mexico. If comparable acreage of the United States were taken by a foreign country, the retaliatory response would be impressive and the cause of the losses eliminated. Furthermore, once delivered for free by floods, the landforms earlier created by the freshly deposited silt served as barriers to increasingly damaging hurricane winds and surf. Today's costly and often counterproductive seawalls and levees fail to keep up, leaving Louisiana in a struggle to stave off oceanic elements in a battle that can never be won but only waged with endless expense. None of this, of course, was ever factored into the costs of the flood-control dams upriver.

END OF AN ERA

For flood control, the Army Corps was authorized by Congress to build one of the most pivotal projects in the history of American dam building.

On the Stanislaus River, New Melones Dam was justified so that low-lying farmland of California's Central Valley would not flood so frequently, thereby enabling farmers to convert from already profitable annual crops and dairy farms to capital-intensive almond orchards and vineyards, all supported by the Valley's congressmen though the project failed to pass economic reviews by state agencies.[27] No average dam, this one rose 625 feet and flooded the West's most popular whitewater rafting river, the deepest limestone canyon on the Pacific coast, and a place of free-flowing wonder that catalyzed America's most intense conflict ever about a dam.[28]

New Melones was ultimately built and filled; however, termination of the Stanislaus as a free-flowing stream represented the last dam fight lost by river conservation advocates at the epic scale—no less than the end of the big-dam-building era in America. It marked the culmination of a citizens' movement that had grown for several decades in sophisticated political dimensions for protection of the finest remaining natural rivers nationwide.[29]

Central to this transition, Brent Blackwelder of the Environmental Policy Center, based in Washington, DC, first became engaged in flood-control debates by exposing shortcomings of the Soil Conservation Service's work channelizing and draining wetlands. Moving up the ladder of damages, he led some of the nation's highest-pitched battles over dams,

The Stanislaus River of California carved the deepest limestone canyon on the Pacific coast and was the most popular whitewater in the West when New Melones Dam was built downstream from this photo, on the left, in 1979.

The 625-foot-tall New Melones Dam flooded the wild Stanislaus canyon in 1980 but marked the end of the big-dam-building era in the United States.

including many being built with intentions to stop floods. A master of logic, analysis, and debate, Blackwelder seemed half a thoughtful and kindly college professor—bespectacled with black-rimmed glasses, always thinking one step ahead of everyone else—and half a cold-blooded political operative capable of altering entire votes of Congress.

His motivation was simply that "special places were being destroyed for no good reason, and at public expense." With that gospel he testified to Congress a hundred times through the 1970s and 80s and became the pivotal point in stopping 140 improperly justified dams and other water developments, mostly by the Army Corps. Keep in mind that by this time virtually all the truly economic dam proposals had already been built. Blackwelder and his own army of sorts bucked all the political odds and blocked federal funding for further dubious developments by halting appropriations bills ten years straight. Playing by the Corps' own rules, Blackwelder's team maintained that the dams' justifications were inflated with "hypothetical" benefits while calculations of costs were "still being swept under the rug of the boosters' momentum."[30]

Among the flood-control dams that failed to pass the 1970s era of new environmental and economic checkpoints, Sprewell Bluff Dam on Georgia's Flint River was halted when Governor Jimmy Carter called out an unjustified economic analysis. At Kentucky's Red River a citizen uprising of Appalachian mountaineers stopped the dam that would have flooded not just farms but also unique forests where sandstone arches ranked second in grandeur only to Arches National Park. At the Delaware—the only main stem river in the East still unimpounded—Tocks Island Dam would have polluted water downstream to Philadelphia with mats of algal eutrophication stemming from the reservoir's heated flatwater. After a spirited citizens' campaign, the Delaware was instead designated a National Wild and Scenic River. Three involved states and local entities pursued other flood-control alternatives, effectively avoiding serious flood damages ever since.

WHEN FLOOD-CONTROL DAMS FAIL TO CONTROL FLOODS

Apart from the problem of dam failures that cause floods of their own (see chapter 2), some dams explicitly intended to control floods fail to do so, and that happens when the dams are needed the most.

In what's widely considered to be a commendable record, dams built for flood control usually stop small and medium-sized floods; however, large floods have repeatedly exceeded the ability of dams to hold back enough water. At that point, uncontrolled flows pour over emergency spillways, and failure to accommodate flows can put the dams at severe risk, as we saw at Oroville in 2017.

Even when a dam is strategically placed to intercept floods, it can fail to do so owing to extreme weather and management protocols. At California's Tuolumne River runoff thundered through Sierra Nevada canyons in January 1997 and poured into the 25-mile-long Don Pedro Reservoir—

large enough to reduce the peaks of high flow downstream. State and federal licenses required management provisions for flood control, but irrigation districts owning the dam responded to a drought-versus-flood dilemma with clear priorities to maximize irrigation, and reserved only 17 percent of reservoir space for floodwater.

People downstream thought the flood issue had surely been addressed with a reservoir that new and large, and it behooved real estate agents in trendy riverfront neighborhoods to not mention floods. As it turned out, a garden variety, 60-year-frequency storm in 1997 quickly filled available reservoir space and spilled 50,000 cubic feet per second downstream, breaking levees and wreaking damage through the city of Modesto. Defending its management policies, the Turlock Irrigation District, as operator of the dam, rationalized to *San Francisco Chronicle* reporters, "You don't want to run into problems during a drought. If you run out of water, that'll impact the whole community, not just the people along the river."[31]

Other floods have failed to be contained when the dams were most needed. A classic case, the Bureau of Reclamation built Pactola Dam in 1956 for flood control, 14 miles upstream from Rapid City, South Dakota. Nearly all the watershed affecting the urban area lay above the dam, well-positioned to catch Rapid Creek's runoff, and a sense of complacency settled over the historically flood-prone town. But on June 9, 1972, a thunderstorm stalled precisely downstream from the dam and upstream from the city. Fourteen inches of torrential rain sent runoff raging through Dark Canyon and Rapid City, killing 238 people.

Increasing the ante of risk in 1997, San Joaquin floodwaters rose within 4 inches of the top of 319-foot-high Friant Dam upstream of Fresno, California, home to half a million people. With more rain, disaster there could have canceled out the flood-control benefits of the dam for the project's entire lifetime.[32]

Close calls keep getting closer in a trajectory that gives pause to anyone looking at where the curve of threats is headed. The Bureau of Reclamation operates two large dams in Idaho's Boise River where three floods in spring 1997 each topped historic crests. Bureau Director John Keys juggled operations skillfully as his agency "raised the river right to flood stage" in the city of Boise, immediately downstream. But Keys modestly added that if another storm had arrived, he "would not have been able to hold it."[33] Then the safety of the dams would require releasing floodwaters on the city rather than risking dam failure.

Hydrologist Philip Williams summarized these dilemmas and risks in *Civil Engineering*. "Many of our huge flood-control dams cannot provide the protection originally promised. Inevitably a major flood will overwhelm the reservoir capacity and break the levees that form the last bastion of defense for the cities and industries that have cropped up on the floodplain, causing immense damage and threatening the lives of thousands of people. People are being deluded into thinking they have a lot of protection."[34]

Harsher critics warned that there are two kinds of flood-control dams: those that have failed to contain the largest floods, and those that will fail in the future.

When Folsom Dam was built in 1956, 30 miles above Sacramento, Bureau of Reclamation planners projected it would protect downstream residents from a 250-year flood. But in the next three decades, eight storms each produced runoff exceeding what the dam was designed to contain. A bit slow on the uptake, the Bureau in 1965 recognized the error and downgraded the dam's protection rating, saying it would protect against a 125-year storm—a calculation still wildly hopeful. In 1986 runoff topped reservoir capacity by a bloating 20 percent. With levees and the state capital at dire risk, city officials feared an "imminent disaster."[35] In the aftermath, safety expectations were again ratcheted down, this time to a 67-year flood, kick-starting reforms. A program of levee reinforcement as well as reprogramming the reservoir's management were again bolstered after a 1997 flood, now making an emergency less likely, and the dam is being raised to better contain floodwaters. But even before the necessary flood-control improvements were completed (still underway at this writing), local land-use regulators and the Federal Emergency Management Agency approved construction of an entirely new wing of Sacramento called Natomas—not just your typical riverfront neighborhood but 90,000 people on floodplains behind levees downstream from the dam. California's Friends of the River warned of flooding threats, but new residents eagerly moved in to the trendy spot, which, with the river out-of-sight, seemed safe. Yet the floods of the future will be higher than those of the past. Even without them, the hazards of levees became apparent with minor flooding in 2023 when Natomas' pumps failed. Except for backup diesel generators, local stormwater would have immersed the new city.[36]

In this inventory of close calls, and some that go beyond close, consider America's largest river. In 2019 the longest siege of recorded floods along the Mississippi shattered expectations for dam safety. After already-saturated soils in the upper basin had frozen, heavy rain pounded the ground, impervious as concrete. A Missouri tributary, the Niobrara River, pushed mega-ton chunks of ice like a wrecked train over Spencer Dam, and downstream the cocktail of solid and liquid runoff filled the 25-mile-long reservoir behind the Army Corps' Gavins Point Dam. Facing a dilemma—to release floodwaters that would engulf farms, homes, and even low-lying towns downstream, or to risk dam failure and a 45-foot wall of water with domino implications the whole way to the Gulf of Mexico—Corps manager John Remus made his best call and dumped 100,000 cubic feet per second from the dam—an equivalent of Niagara Falls that no one ever expected to see in Nebraska. Remus humbly reflected, "It's human nature to think we are masters of our environment. But there are limits."

Vic Baker, head of the Department of Water Resources at the University of Arizona, reflected broadly on this flood-control problem of the ages and pointed out, "Dams provide protection against little floods. But when there is a really big flood, the high priority is to protect the dam."[37]

Even without structural shortcomings such as those that Remus and Baker faced and feared, the chance of a dam being inadequate increases as the structural reliability of this critical infrastructure ages and as the intensity of flooding increases. Dams may be ineffective or counterproductive when people need them the most, and a false sense of security can be worse than no security at all.

BEYOND DAMS

So, back to my question at this chapter's opening: who can argue with the logic behind building a dam for flood control? Well, a lot of people can.

In 1936 the promise of damming was clear. While the logic was not wrong, it was, at best, incomplete and imperfect, and at worst it led to delusions, waste, and counterproductivity. With a timeless perspective back in

1954, the famed hydrologist Luna Leopold wrote, "Flood control does not mean the elimination of floods. At best, it can provide only a certain amount of protection against overbank flows."[38] In what may be the worst possible outcome of diligent, honest, and expensive efforts to stop floods, the "good years" of being spared smaller floods have enticed people to invest more heavily in property that will ultimately be flooded during the "bad years." This outcome now grows more likely under an ominous triple-threat and "perfect storm" of error: floods are growing larger, dams are growing older and less secure, and floodplains are growing more populated.

Per capita losses from floods more than doubled between World War II and the early 1990s in spite of that period being the heyday of flood-control-dam building.[39] Investment in dams and levees exceeded $25 billion as of 1992, yet damages increased to more than $2 billion every year according to the Federal Interagency Floodplain Management Task Force in 1993, and after that the damages continued to increase.[40]

Lessons from the history of dams remind us that no response relying entirely on structures, engineering, variable maintenance, and incomplete knowledge of fundamentals as basic as the weather forecast can be failproof. Through this painful history of gains and losses, it has become evident that the flood-control dams of the past are not adequate to protect people, farms, and cities today, let alone tomorrow.

The knee-jerk response to this reality—build more dams—does not appear to be the answer. It's widely believed that virtually all sites considered feasible from a hydrologic and economic standpoint have been built upon. As Brent Blackwelder said forty years after he led a movement to halt the nation's habit of doubling down on flood-control dams whether they were economic or not, "Even with greater storms, we're not going to see any significant rekindling of the dam-building era. The feasible sites are already dammed or off limits for other reasons." Bearing this out, the federal dam-building agencies have not offered any serious proposals for new large flood-control dams in several decades, all in spite of the floods getting worse. Flirtations with proposals end with abrupt rebuke when cost estimates—such as those for St. Petersburg Dam, considered for Pennsylvania's Clarion River in the wake of Hurricane Agnes—show nothing but red ink.

Pointing out the dearth of feasible dam sites remaining, and pressing for a no-nonsense grip on the prospects ahead, Jay Lund, co-director of the Center for Watershed Sciences at the University of California, reminded us, "There are 1,500 reservoirs in California right now. If you build a new one, you will be building it in the fifteen-hundred-and-first-best location."[41] We've run out of dam-building options, and modern standards for safety, environmental trade-offs, social acceptance, and economic calculations encourage and mandate more selectivity, not less, about where we build dams.

The past pattern of overestimating security provided by the dams argues against depending on yet more of them. Further, and sealing the fate of ambitions to build new dams in the future, every penny one might imagine spending on dam construction—and then considerably more—will be urgently needed to simply maintain and rebuild for nominal safety or utility the 400 flood-control dams plus other public works that we already have. Just consider the $4 billion upgrade to Sacramento's already-expensive flood-control system. Think about the reinforcements for safety at Wolf Creek Dam on the Cumberland River in Kentucky upstream of Nashville, Tennessee, that cost far more than did the original dam. And bear in mind the new spillways at Oroville Dam that cost ten times the original contract for dam construction. These and other upgrades have not been optional, but mandatory, busting the bank for whatever might otherwise have been budgeted to build new dams.

Which brings us to a troubling future. If the dams' safety and reliability have been questioned in the past, they will be increasingly challenged during the intensified floods that a stormier climate promises.

Furthermore, few discuss it, but the ultimate fate of this entire form of flood control is yet more vexing. An effective taboo surrounds this topic, but every dam's days are numbered. Some will fail because of inadequate maintenance and unforeseen hazards. Some will be overtopped and debilitated by floods of unexpected size. Some will suffer fractures by earthquakes, which is why Auburn Dam was halted in California even after $200 million was spent. In the grand scheme of time, dams are victims to ultimate inevitabilities beyond anyone's control. Cracks will riddle the aging structures just like they do the cement of an old sidewalk, though unlike fractured pavement underfoot that might cause us to stumble, disintegrating dams can bring whole cities to their knees.

Finally, consider this: if dams don't succumb to other failings sooner, all of them will fill with silt and become useless for flood control for that reason alone. It's not "if," but "when," and "when" may not be that far off. A United Nations report in 2023 found that dams in the US will likely lose 34 percent of their cumulative flood-storage capacity by 2050.[42] Acknowledging the temporary nature of benefits even from Glen Canyon Dam on the Colorado River, former secretary of the Department of the Interior Stewart Udall said, "It's going to fill up with silt in a hundred years, and what is the judgment of posterity then? That's the real down side of the big dams—the rivers are not renewable."[43] Regarding that fate, former commissioner of the Bureau of Reclamation Daniel Beard acknowledged, "This is one of the great unknowns in the water resources business. . . . It is clear that every dam you construct is eventually going to silt-up. One of the great myths about dams is they are here for eternity—that they will function in perpetuity. And, of course, they won't."[44]

Not unlike passing the buck on nuclear waste that will plague future generations for centuries, and centuries, in exchange for a few decades of relatively cheap electricity today, the ultimate disposition of the dams is a taxing and dangerous liability that we're selfishly passing on to those coming after us. Or, in other words, to our children.

These outcomes are grim, but there is a better way. In 1997 hydrologist Luna Leopold reflected on our dam-building response to floods. "We have done it in a way that was bound to fail. When you rely on structural works, such as dams and levees financed by the federal government, rather than saying, 'We will control land use,' then you are bound to fail. We're attacking the problem of floods when we should be attacking the problem of land use."[45]

The heyday of dam building in America—1936 to 1980—is past. The time has come for other solutions, as Dr. Leopold sagely suggested. Even if the dams have helped, rather than worsened our hazards and vulnerabilities, they do not protect us from the floods of the future.

But before we see how to do that, let's recognize that building dams was only half of our approach to flood problems in the past.

5 Broken Barriers

IN LOVE WITH LEVEES

For tens, hundreds, and thousands of miles along the Mississippi and other rivers, America's farmers, cities, levee districts, and the Army Corps of Engineers have constructed levees in hopes of containing floods and keeping them out of the places where we live and work.

Building levees, even more than dams, represents America's most common and widespread response to the hazards of flooding and is, in many cases, the infrastructure of choice. Levees are easier to build than dams and can be thrown up, in one form or another, almost anywhere in the risky zone between house and river. Some levees are massive bulwarks such as those outlining Mississippi shorelines like paralleling freeways on each side, as long as the continent is wide and visible from space. Some are barely noticed in tiny burgs such as Hyndman, Pennsylvania, where a pile of rock-and-dirt only a third of a mile long confines Wills Creek to its narrowest channel.

It's sensible to try to block out any intruder who might threaten our property. Build a wall, fence, or gate. But floods are liquid culprits and cannot be barred so easily. Before relaxing in the comfort zone of a levee, even if engineered to compulsively safe standards, think, and imagine a window, like a sealed porthole in the side of a ship, but located within the levee's thin width. During a flood, what you would see through that window is opaque brown water rushing past on the other side at heights towering high over your head. In fact, as much as 30 feet above it, and with the force of a rapidly moving ocean, in some cases freighting runoff from a dozen states. If that vision of floodwater—boiling, angry, and perched high above and directly next to you—seems a bit chancy, it may be.

Levees date to Native Americans' mysterious mounds at Cahokia and to the earliest colonial settlements when residents cobbled up a berm around houses to keep water out, or at least to try. This likely happened at the first two American towns ever built by Europeans: the Spanish outpost of San Juan de los Caballeros along New Mexico's Rio Chama in 1598, and nine years later at English pioneers' ill-fated Jamestown in Virginia's riverfront low-country. The first major levee was perched above the Mississippi in 1717 at New Orleans, a structure repeatedly augmented and rebuilt in aggrandizing patterns continuing to this day.

In 1850 Sacramento voters approved local funding to erect a 3-foot-high levee, obviously inadequate in retrospect. Legislation soon established state policy of converting "swamplands" into farmland by ditching, draining, and leveeing wherever possible. Thousands of escalating efforts followed nationwide. Along the lower Mississippi, 1,000 miles of private levees—pick-and-shovel work by slave or virtually slave labor—appeared by 1860. Floods routinely overtopped them, followed by reconstruction to a higher level, up, up, up. Still the waters rose higher than the height of the levees in the riparian zone of escalating warfare against floods.[1]

The federal government weighed in with its first authorizations for flood control in 1917, greenlighting not the flood-control dams with their gracefully arcing concrete faces that later symbolized the Army Corps of Engineers, but the more mundane piles of rock and soil constituting levees or, as some people call them, dikes. Until 1936 the Corps' flood-control efforts supported not just more levees, but *only* levees, including a nationalized effort to discipline the wandering, multislough Sacramento River into a single obedient channel draining the bulk of runoff from a dozen major rivers, even though state agencies disagreed with the approach.

More remarkable was the Corps' confidence in trying to keep the entire Mississippi at bay. Go figure: our largest river stretches 2,340 miles and drains 40 percent of the US outside Alaska. Levee builders with their picks and shovels and later their backhoes and bulldozers were up against an inland sea. How could they not know they were outgunned? But doubling down whenever its past efforts failed, the Corps took greater and greater steps into deeper and deeper waters with the "Edgar Jadwin Plan" of levees, levees, and more levees, a strategy named after the general who led the agency in raising and strengthening hundreds of broken and washed-out bulwarks following disastrous 1927 floods.

In spite of failures, belief ran deep that private property could be saved by levees, and the extremes of that belief extended to paranoia and ostracism aimed at anyone who disagreed. After sensibly questioning widespread reliance on levees in 1939, geographer Gilbert White was personally investigated by a House of Representatives committee that perceived his caution as potentially "un-American."[2]

Backed by a federal government so convincingly entrenched and numb to other options, or for that matter to other thoughts such as those of Dr. White, people had few compunctions about investing in homes on flood-prone ground sitting behind the levees, and farmers working there shed their formerly hardwired fear of losing crops. Along the Missouri River the Corps built hundreds of miles of levees, and on the dry side "new" land was offered to farmers, who jumped for it at low cost once it was "protected." Corps records reveal that 522,000 acres of floodplain there were converted to farms in the levee-building spree of the twentieth century.[3] This transformation of America's greatest bottomland forests into agribusiness monocultures occurred while taxpayers footed most or all the levees' costs.[4] Just like flood-control dams, levees were believed to make floodplains safe, productive, and profitable.

The numbers and mileage of levees are unknown. But the largest, most formidable, and costliest total 33,000 miles—San Francisco to New York, not once but thirteen times. The Army Corps is associated with 13,412 of those miles; the rest fall to other jurisdictions.[5] The Federal Emergency Management Agency (FEMA) and other federal agencies and states have registered another 99,000 miles of typically smaller levees.[6] Thousands of even lesser ones can be found where all it takes is a bulldozer at waterside and compliance with a federal dredge-and-fill permit or—more likely in many local cases, including some I've seen while canoeing in streams scattered across rural America—a willingness to quietly work on a homemade levee at the "back forty" and not tell anyone.

Levees principally do one of two things. They protect farms, which account for most of the levees' linear mileage nationwide, and likewise for most of the floodplain acreage lying in the "protected" zone behind the levees. Second, levees safeguard urban areas, including downtown St. Louis, Memphis, Omaha, and Portland. Cities and towns account, by far, for the largest population and highest monetary values guarded by levees. Levees in urban areas are typically built by the Army Corps of Engineers, though they account for only 18 percent of the total built by that federal agency.[7] Keep in mind, also, that the mileage of urban levees is much lower than that 18 percent when all jurisdictions are counted, as many of the levee builders are explicitly agricultural organizations. The vast majority of total levee mileage protects farmland with its principal crops of corn and soybeans in the Midwest. In California 1,700 miles of levees guard farmland in the 400-mile-long Central Valley, while only 200 miles, or 12 percent, safeguard urban or housing acreage, even in the West's most urbanized state.

According to the Army Corps' National Levee Database, 16 million residents and 5 million buildings are located in areas that would otherwise flood but are situated behind levees. But note: those numbers include substantial populations along seacoasts and estuaries rather than rivers. FEMA estimated that federal levees nationwide protect 56,000 square miles—the size of Iowa and mostly farmland.[8] The take-home fact stemming from these statistics is that relatively few miles of levees provide the lion's share of benefits, and that the vast majority of those economic benefits occur in urban areas.

This lopsided investment in protection of sparsely populated rural floodplains owes to federal policy whereby for decades America's taxpayers funded 100 percent of levee construction, no matter how local or how few the beneficiaries. This continued until the Reagan administration in 1986 required cost-sharing, which still left American taxpayers with a heavy lift of 65 percent. In spite of the cost-share rule, reconstructions and major repairs—frequently exceeding the levees' original costs and often doing so by enormous sums—are often covered principally by the federal government. Fixing broken levees is universally considered a top priority for public spending, and certainly so compared to efforts to help relocate both urban and agricultural investments off of floodplains.

Levee defenses are often hidden behind backyards of neighbors or on the other side of streets where few people go, even in a city as large and levee-dependent as Sacramento. Because of that, many people don't realize they'll be flooded if and when their levee fails or when floodwater overtops it.[9]

Uncounted or partially omitted in calculations of benefits and costs, levees affect fish, wildlife, groundwater, and the health of river systems. The levees block nourishing flows from reaching riparian habitat—richest of all for wildlife. They channelize a river's current, and in the process they destroy streambed complexity, preempt the pool-and-riffle sequence, and eliminate natural shorelines. They prevent groundwater from soaking into floodplains where it can nourish streamflows later in the year. They cause erosion of river channels, or under perverse conditions of hydrology they cause the opposite: aggradation whereby silt and gravel build up directly in the channel between the levees, making the riverbed higher. This smothers bottomlife that's critical in nature's food chain, not to mention that it forces flood flows up higher against the damage-prone sides of the levees and ultimately over the tops of them. Dr. Jeffrey Mount of the University of California at Davis summarized the problems with levees, saying, "There is no other engineering design, perhaps other than dams, that more directly contravenes the natural processes of a river than a levee. It divorces a river from its floodplain. One of the most dynamic systems in the world is asked to hold still when you place levees on it."[10]

Levees' shortcomings are generally tolerated in urban areas, and understandably so with a lot of dollars and infrastructure at stake. But beyond the cities and towns, the environmental downsides raise harder questions of spending, of collateral damage, and of alternatives to reduce flood damage.

Most important to many people, a levee in one place pushes flood flows higher elsewhere, which can make calculations of net gains questionable. Even more important to many is the increasing frequency and intensity of storms that can render levees useless or worse.

LIVING ON THE EDGE

For a year I lived in a pleasantly shaded neighborhood protected by a levee in Sacramento, spending days and nights below the flood level of the American River but not worrying or even thinking much about it. Near as I knew, nobody did. On summer afternoons I rode my bike a dozen blocks, then walked up the levee's steep slope, 30 feet high, and down the other side to a beach, delectable with chilling currents and a happy scene brightly animated by swimmers and sunbathers.

But on a rainy February 20, 1986, I climbed to the top of that levee and found a flood swirling on the other side. Waves lapped at my feet like a Pacific surf pounding at city's edge, except this was 75 miles from the ocean. Crusted enough for a pickup truck, a veneer of gravel on top of the levee felt solid underfoot, though I couldn't see what was happening at the levee's core or on the water-swept slope angling sharply into the river, which carried runoff from everyplace between there and the wintery 10,000-foot crest of the Sierra Nevada range whose combined snowmelt and rain whisked down to sea level just south of the city.

Miles from the forests capable of producing them, logs tumbled within the chaos of frightening chocolate water. In my view the other way, across the local streetscape, Sacramento looked like a toy city of Plasticville roofs. I knew that thousands of homes lay beyond what I could see, including the house where I was staying, all huddled below river level that day, some a full two stories beneath the surface of the angry flow. I happened to be witnessing a flood thus far contained by levees but pushed up to unprecedented heights because of them.

That mind-opening and quizzically disturbing view to a city situated below the breaking waves of a raging river—wide as an arm of an ocean— sparked my curiosity to see more, so I bicycled upstream on the levee-top trail until I came to a bizarre sinkhole. Called "sand boils" in flood-fighter lexicon, these form on the landward side of levees and are caused by water infiltrating the earthen structure. While sand was absent in this case, the "boil" aspect of the name was clearly evident in bubbles and burps of water within the hole, looking like an artesian spring. City workers had placed sandbags around the eroding cavity in an effort to reinforce that precariously thin separation between city and river. But then the defect, like an

In 1986, flood flows of the American River undercut this riverfront bicycle trail in Sacramento and approached the tops of levees protecting California's capital city. The same flood breached similar defenses in nearby communities and led to expenditures of many billions of dollars in reinforced levees within urban areas.

advanced cavity in the outsized teeth of the city, seemed to have been forgotten. Nobody stood there monitoring it. Truth be known—later—what I saw was a mere Band-Aid on a gaping wound. Water in the bagged sinkhole slightly rose and fell with subterranean pressure, indicating, I presumed, variable forces in the surging river on the other side. The revealing window to muddy flows that I've suggested we imagine through the face of a levee almost existed, but rather than being closed and caulked like a porthole in a ship, it was open to reaming by a potentially unstoppable jet of water coming from the wet side.

Another day of rain would have pushed water levels even higher because the reservoir behind Folsom Dam, 30 miles upstream, had swollen to brimfull, requiring the federal Bureau of Reclamation to release everything flowing down from the snow-covered and now rain-sodden mountains 60 miles eastward. The bloated reservoir created the troubling specter of the dam's collapse, which would result in all of Folsom's water cascading at once. Though catastrophic, levee failure is a far better option than dam failure, which is guaranteed to cause the levees to fail as well. Walking a tightrope of management, flood officials dialed both dam and levee brimfull in a perspiring game of extremely high stakes, all homes, businesses, and residents reduced to chips on a gambler's table. Had I realized the magnitude of weakness that I had stumbled across, I would not have spent the day sightseeing high water by bike but rather fleeing to a safe distance on higher ground.

The rain stopped, the runoff crested, the levee held. But nearby communities were less lucky, and "luck" is an appropriate description of Sacramento's fate at that time. Just northward, in the city of Linda, a Feather River levee ruptured and sent 26,000 people fleeing homes and shopping malls, all having presumed they were safe on the dry side of public works.

Meanwhile in the state capital, scarcely anybody knew about their personal brush with destiny, cryptically evident to me as I stared at the belching sand boil that secretly and subtly threatened the city's ramparts. Half a million of us survived that night in unimagined peril. Thirty years later, Rick Johnson, executive director of the Sacramento Area Flood Control Agency as of 2010, reflected in a *Sacramento Bee* interview, "The scary part is you couldn't see the damage to the levee. It was all underwater. We didn't even know that was happening until after the water came down. They should have evacuated, quite frankly."[11]

In the retrospective wake of that flood, California's capital city earned dubious distinction as America's "most-flood-endangered city," and tough competition for that heavyweight title included even New Orleans.[12] Rising late but admirably to the challenge, local leaders formed the Sacramento Area Flood Control Agency in 1989 to establish better protection. Substantial upgrades were followed by a secondary round of investments after another close-call of flooding in 1997. A medley of strategic, commonsense, and creative solutions included inserting impervious interior walls into the levees to halt the kind of seepage and sinkholes I had seen, raising levees higher, hardening their exterior slopes on the wear-and-tear line confronting the river's currents and abrasions from the multiton logs I had seen, retrofitting Folsom Dam to better release floodwater, revising the dam's operation plan to balance the interplay of dam-and-levee dependence, paying hydropower dam owners farther upstream to dedicate space for floodwater when push came to shove, and upgrading a downstream weir to flush runoff onward and avoid the reflux from even greater flows in the Sacramento River, which the American River joins within the city. None of this was cheap. Improvements since 1986 will total $4 billion or more when complete in 2026. Anticipating growing risks with global warming, the Agency aimed to accommodate twice the volume of the greatest flood in the past.[13]

Sacramento might now be regarded as a model of involvement and investment necessary for safety behind a levee. Also an example of the cost—the truly astronomical cost—required to actually protect a low-lying area from floods. It's also emblematic of the degree of risk that people can unknowingly face when the care and feeding of a structure as mundane as a levee is not regarded seriously. It's no small undertaking and, let's face it, an impossible one for most communities lacking the robust population, the professional municipal expertise, and the bold economic profile of the capital city in a state that, all alone, boasts the world's fourth-largest economy when compared to 195 nations.

Sacramento's Cadillac levee is emphatically not attainable for most communities and most acreage now located behind the 33,000 miles of major levees nationwide, let alone those hiding behind tens of thousands of miles of smaller and less reputable fortifications.

Circumstances beyond Sacramento are more representative of the state and the nation, where many levees are insecure, underfunded, and overly prone to failure. Look no farther than the rest of the Sacramento River basin, where thirty levee breaches occurred in 1997 alone. California's Department of Water Resources reported in 2011—a full quarter century after the 1986 scare and its motivation to arm up in flood control—that half the miles of urban levees statewide still didn't pass basic engineering standards. And that cohort of urban levees happened to be the "good" ones. Most of the state's levees are, instead, definitively rural, 60 percent of which flunked inspection and scored an alarming "high potential" for failure, an expectation no one wants to sleep on.[14]

Look, also, no farther than the neighboring river basin. In a mirror-image of the Sacramento River's southbound flow from its headwaters, the northbound San Joaquin River confronts head-on the larger Sacramento in America's most expansive inland delta. Imagine two dueling garden hoses precisely on-target with each other and the splatter of water that would ensue. At the edge of this imposing geographic feature, promising a significant amount of hydrologic chaos, lies the levee-dependent city of Stockton, population 311,000 and growing, most people seemingly as oblivious as I had been to flood hazards while living among them. Stockton's levees precariously guard against increasing floods of the San Joaquin and also the perverse southward push of the far larger Sacramento River, not to mention the oceanic estuary that's rising eastward toward the city's lowly elevation, 13 feet above sea level—not much higher above the Pacific Ocean than a basketball hoop. That's an unimpressive safety margin considering any one of the following: storm surges in the ocean, tsunami crests from earthquakes as distant as Japan, floods barreling down not just the arterial San Joaquin and Sacramento Rivers but also the winding webworks of the Mokelumne and Cosumnes Rivers that empty into the delta between the two larger streams, and topping it all off, a legendarily seismic substrate of geology where a single earthquake tremor could cause multiple levee breaches cracking open within minutes of each other. No, make that within seconds of each other. All this is unsettling and even more unnerving considering that an atmospheric river carrying a 1-in-200-year-or-worse-frequency storm is due, and much of that water will funnel from all directions down to Stockton, the clock ticking in this "surround-sound" version of hydrologic threat.[15] Floods in this city are going to be a problem. With increasing dangers elsewhere, other towns nationwide and most of our rural floodplain mileage lying behind levees wander adrift in similar leaky boats.

HIDDEN DANGERS

Like hundreds of thousands of other people living squarely on Sacramento's urban floodplain, I had casually taken my security for granted. Untold millions do the same in cities and countrysides across America where levees have proven to be safe, except where they haven't, and most personal and community decisions about land use in the levee zones are based on the "safe" part of these conflicting truths.

Here's the difference between them: the levees are safe in minor and midlevel floods, but many fail or get overtopped in large floods. Of course, large floods are the ones that count the most. Large floods, by the way, are getting larger, occurring more often, and arriving more suddenly (see chapter 6).

Add to apprehensions, here, that levee building is a zero-sum game. If only one side of a river is protected by a levee, it pushes floodwaters to the other side, and with gusto because the overflow there is doubled in volume and accelerated in speed, all raising issues of equity, cost, and security. Renowned biologist and environmental movement pioneer Barry Commoner's second law of ecology confidently states that everything has to go *some*place, and the repercussions of that self-evident truth trouble levee-guarded societies to their vulnerable cores. Meanwhile, they trouble the neighbors of those societies even more.[16]

So, to avoid worse flooding on the opposite shore, we build a levee there too. But that's no get-out-of-jail-for-free card. Levees on both sides create a bottleneck that speeds the flow and forces it higher between the levees. This not only strains structural integrity of those bulldozed defenses through added stress on both sides of the river, but also causes back-pressure and aggravated flooding upstream because the levees pinch the river from both sides and force incoming water to back up and wait in line, so to speak. Levees also cause greater flooding below, where the concentrated jet of runoff suddenly escapes bondage and spews out widely and uncontrollably again, like air spitting forcefully out of a balloon that you blow up but then let go without tying the knot. The resulting flood reaches even farther now because the levee above has displaced the water's spread there. Typically, communities both above and below the levee districts are, well, simply out of luck, with no recourse in their victimization by the levees that have been built by more populated, affluent, and politically savvy communities both up- and downstream.

Towns have engineered levees to protect business districts, sometimes leaving low-lying neighborhoods—especially those of the poor and politically powerless—not "out to dry" as the cliché might symbolically imply, but literally out to flood. Look no further than Mark Twain's picket-fence Americana of Hannibal, Missouri, where a flood in 1973 motivated local officials to proceed with a project the Army Corps had proposed. But to save dollars in the local cost-share, the city shortened the levee, leaving the lowest-income people—mostly African Americans—unprotected in the shrunken outcome.[17] I can just hear Huck Finn's Black pal, Jim, patiently pointing out the inequities in a modern rewrite of Mark Twain's classic. Those inequities often lead not to solving people's vulnerabilities to flooding, but rather to building more, larger, and taller levees in more and more places, and it would behoove us, as responsible people, to forecast where this scenario ultimately takes us.

The newly jeopardized towns or farm districts beyond a levee's reach often try to build a levee of their own, overspending themselves badly in what often proves to be futile effort, and on and on it goes, upriver and down in an escalating war on floods and inadvertently a war on each other, all having no end or exit strategy. The more we constrict the river, the more it rises, speeds up, tears at its banks, backs up, and overtops the levees meant to contain it. Ultimately it's difficult for anyone to win this bankrupting battle, and it's starkly impossible for *every*one to win.

Human nature and economics being what they are, some towns and farmers have raised their levees to be explicitly higher than those of towns and farms on the other side, triggering undeclared "levee wars." At times these have begun to resemble real wars, infamously conducted by armed possies, both sanctioned and mobocratic. Levee construction or heightening can become blatant aggressions against otherwise friendly forces— towns where your own aunt, uncle, and cousins might live just across the bridge. Historian Robert Kelley documented these competing levee efforts involving "neighbor against neighbor, upstream vs. downstream, farmer against miner, one political party against another."[18]

Unlike in early years when the Midwest acted like the Wild West with shoot-to-kill vigilantes guarding the home turf of local dikes along the Mississippi, regulations now control heights of levees on major rivers. But one has to wonder about the regulators' efficacy when some flood districts still build beyond authorized limits, or when they still act with understandable but illegal abandon by sandbagging higher during the immediate panic and passion induced by cresting floods and their roar that drowns out other voices, including those of both neighbors and the law.[19]

The Wild West is gone, but people still struggle with levee wars. An east-side agricultural district has elevated its Mississippi River levee to spill flood flows onto west-side farms and into the town of Hannibal, which, as we've seen, has suffered its own issues of fairness in levee matters. A bit late, it seems, the Army Corps ruled that the 60-mile-long farmland levee had been extended 4 feet higher than permitted, according to a *St. Louis Post-Dispatch* investigation in 2015.[20] Recognizing more than an errant violation slipping through administrative cracks, Nicholas Pinter, formerly of nearby Southern Illinois University, pointed out, "The broader thing here is that a lot of these levee districts manage their floodplains with impunity." In 2023, during the remarkable siege of California runoff that poured into the long-dried-up bed of Tulare Lake, reports circulated of self-interested backhoe operators deliberately breaching levees to protect their own property at the expense of others.[21]

Those are explicit battlegrounds in levee-land's trench warfare, but even where two sides of a river fight to a draw, with levees at a uniform height, trade-offs come into play, and one of these balances the height of levee protection against the danger of levee failure. What is noticed by anyone tightening the nozzle on a garden hose, and in that way creating greater pressure and force in the water squirting out, was brilliantly quantified at the scale of America's largest river by St. Louis University geologist Charles Belt with a landmark article in *Science*. Levee constrictions created all-time record crests at St. Louis in 1973. In spite of that flood being only of 30-year frequency carrying identical volume to a flood in 1908, it climbed 8 feet higher against the city's levee.[22] Other analysts have concluded that many parts of the levee-girded Mississippi now top out 10 feet higher than they did with flood volume comparable to the historic crest of 1927.[23] The higher the river rises against the levees, the more likely the levees are to fail.

Knowing all this professionally, and considering a proposed levee at the confluence of the Missouri and Mississippi just upstream from St. Louis, retired Army Corps Colonel Edwin Decker colorfully assessed, "You have to be out of your cotton-pickin' minds to even consider building a levee at the confluence of these two rivers." With an eye to nearby farmland as well as the city of St. Louis, now housing 305,000 people plus substantial suburbs, Decker added, "Nature developed the floodplains as a safety device, and now you want to take that away."[24]

Addressing the issue more recently and broadly, Robert Criss, hydrogeology professor at Washington University in St. Louis, noted that Mississippi flood levels reached 40 feet only once per 32 years on average before 2013 but a shocking once per 1.5 years following that. "For us to be setting a record flood again, this is just absurd," Criss said, blaming the problem squarely on the confinement of levees. "We've messed with the rivers too much, and we've got to lower some of these levees to allow the river to spread out." Regarding critics of this view, Criss added, "A few old river cities, sure, they need their floodwalls. I get that. But trying to wall off both sides of the river for a thousand miles—how idiotic."[25]

Studying multiple Mississippi floods, Nicholas Pinter found that in all cases investigated, floods of comparable volume pushed postlevee water heights higher than those predating the levees, an effect that now has its own anthropocentric name: "magnification of floods."[26] The bottom line is that levees have made flood heights worse. One among many ironies of that fact is that it cost a lot of taxpayer money to do it.

Similar in some ways to levees, navigation structures called wing dams likewise concentrate, raise, and speed up flood flows on major rivers. Far more ubiquitous than people other than a select club of corporate barge operators realize, thousands of these single-purpose structures built by the Army Corps and funded by taxpayers angle out from shorelines in order to concentrate currents and maintain depths of shipping channels, purposefully causing faster flows but inadvertently overflows, shoreline damage, and chain reactions of hydrologic artifice.[27] Affecting not only the navigational workhorse Mississippi, wing dams have also proliferated elsewhere including in the Pacific Northwest's Columbia, with 233 of the troubling encroachments along 145 miles of river. Built at behest of the barging industry and at public expense, those wing dams deter, confuse, and kill fish on essential spawning migrations. Salmon on the endangered species list face the troubling wing dams at a rate of 1.6 per mile.

Channelization also hastens the speed of current and the height of floodwaters. That, in fact, is the purpose of channelization. In this ubiquitous but flawed model of flood management, streams with irregular edges, friction-inducing forests, intermittent shallows, and meandering bends responding to the natural dynamics of flow are straightened, deepened, simplified, and often veneered with concrete in order to sluice water through faster. In other words, an intricate and organic manifestation of nature is converted into an outsized fire-hose aimed at someone else's town or farm directly below. Furthermore, under the immutable laws of physics, the hydraulic pressure resulting from channelization increases not by linear factors but by surprising geometric ones; a doubling of velocity in channelized rivers quadruples the water's erosive force and increases by a shocking sixty-four times its capacity to scour the riverbed. Unseen but real, these dynamics breed cyclones of new problems, not the least of which are flood flows that rip apart riverbanks downstream of the channelizations and that destroy riverside levees. Entangled issues arise about who benefits, who pays, who's a victim, and who's a perpetrator of what could be considered a crime.

Not counting river frontage constrained by levees, channelization affects tens of thousands of miles of rivers, with many of the "improvements" undermining habitat, wildlife, recreation, and fundamental ecological services the streams once provided.[28] A landmark Arthur D. Little study in 1973 estimated that 200,000 miles of stream channels in the US had been modified in damaging ways. For barges alone, 26,000 miles were channelized at a 100 percent subsidy to the barging industry.[29] At a smaller scale, the Soil Conservation Service channelized 21,000 miles of streams—ironically priming them not for soil conservation by the agency so named, but for soil erosion as wetlands were drained and flow velocities accelerated. Legal challenges eventually halted the practice and the SCS name was changed to Natural Resources Conservation Service.[30]

Like all infrastructure, concrete channelizations require maintenance. This is seldom budgeted or even thought about beyond initial federal subsidies to build the channels—a taxpayer gift embracing the darkly cynical refrain "The first one's free." Many of those channels were excavated and paved during the mid-twentieth century, which means they're as badly in need of repair or replacement as is a highway built in the Truman administration and accumulating potholes unpatched ever since. Near San Francisco, Contra Costa County alone needs channel repairs costing $2.5 billion, typical of many communities, though no one can pay that kind of money.[31] Related to such budget woes, the California Department of Water Resources estimated in 2022 that $30 billion are needed over the next thirty years for repairs and upgrades to existing levees in the state's Central Valley—call it one billion per year—yet in a recent fifteen-year period only $3.5 billion were spent.[32] Across America few jurisdictions can keep up with the maintenance needs of a vast infrastructure intended to stop floods but now facing deteriorating abilities to do so, even assuming that the original approach had not been flawed and deceptively justified.

Levees and their kin in the water management business have their problems, and the largest one is this: like many dams that were intended to stop floods, most levees work well during small floods when the stress of the water's force and the height of its rise are not great. But the larger the flood in a leveed river, the greater the water's height, the more unhinging the stress, and the higher the likelihood of runaway flows overtopping or eroding the levee's face and causing catastrophic failure.

Levees have undoubtedly prevented a lot of flood losses. Crunching the Corps' numbers, one might be tempted to say that levees work, which is true, except when they don't, and that's when the floods and the dangers are greatest.

WHEN LEVEES DON'T WORK

The fear of levee failure might strike the public works pragmatist as irrational. Competent engineers design levees, and a lot of money is spent to make them safe. Some of us see levees every day, doing their job, and their familiarity breeds a "no worries" sense of security if not complacency. Municipal professionals, if not the Army Corps of Engineers, are in charge here. But those comforts are not always justified. So consider what happens when a levee breaks. That alarming vision becomes real in newscasts of mud-smeared emergency crews and volunteers battling rainstorms to shore up crippled defenses with sandbags. We love levees when they hold, but it's a real mess when they don't.

The greatest losses in the notorious Hurricane Agnes Flood of 1972 were not along thousands of miles of overflowing rivers, but at a disastrous few nick-points totaling only a few hundred feet in levees intended to protect Wilkes-Barre, Pennsylvania—sister city to President Joe Biden's better-known Scranton. At critical spots the levees had slumped because coal mines had previously been tunneled directly beneath, and like a heat-seeking missile, the flood found those spots, overtopped them, and quickly grew as water stormed through the breach. It's hard to know who to blame: an unregulated coal industry that externalized its costs to later generations, negligent levee maintenance by the city, or inadequate oversight by the Corps. But what we do know is that the failures displaced 100,000 people, necessitating one of the largest such evacuations ever. In a matter of a few hours, or less, the mine-induced, levee-failure overflows caused 40 percent of total damages from the entire Agnes Flood's epic eight-state, weeklong siege—the most costly flood disaster in history up to that time.[33] One might be tempted to dismiss Wilkes-Barre and its coal-mining legacy as a special and unfortunate quirk of the past, but as it turns out, failures for a wide range of reasons have occurred again and again where people have bet their fortunes and lives on levees.

As at Wilkes-Barre, many failures begin when floods overflow the top of the levee. At that point most of the additional rise of runoff fails to escape downstream and instead spills through the levee's low spot and into neighborhoods or fields on the "dry" side. Worse, the overflowing river's erosive power surgically targets the lowest dip in the levee, gouging turf, emulsifying soil, then V-notching downward, quickly overcoming any gravitas that the levee's structural core might have had. Fully breached, the levee behaves like a failed dam and unleashes a wall of water to the side where people live.

Many levees fail even without being overtopped.[34] The sinkholes I saw threatening Sacramento in 1986 could have transmitted increasing flows and led to startling blowouts and full-on levee collapse in America's thirty-fifth largest urban area. A nationwide survey revealed that among 100 recent levee breaches, 80 percent owed to "geotechnical" failures—principally water seepage that begins eroding slowly but escalates out of control and tunnels the whole way through.[35] Fixing levee weak spots—and whole networks of them—has required excavating a 4-foot-wide trench as deep as 120 feet laterally within the center of the levee and filling it with impermeable bentonite or cement slurry. Repairs typically cost more than the original levees did—way more—with expenses never factored into the projects' original benefit-cost ratios, which frequently justified construction by narrow margins.

Problems can start small, invisibly, and from ubiquitous sources that—until you know—seem laughable. Take gophers. These diminutive bucktoothed rodents are a minor nuisance when tunneling beneath the lawn and stacking their conspicuous cones of dirt on the grass, but the industrious little excavators collectively move a lot of soil, and along with burrows of mice, groundsquirrels, and rats, they've given hundreds of miles of California levees a wicked case of Swiss cheese. The tunnels become conduits for leakage that can infiltrate the entire way through a levee, setting up erosive flow-through forces that hollow out the levee like progressively larger drill bits might do, and they end up causing serious failures.[36] Making the gophers look cute, consider the 20-pound, chisel-toothed nutria. This exotic rodent was carelessly imported for the fur trade because its tanned pelts easily pass for mink. But multitudes have escaped their animal-farm pens and have ferally infested riverfronts, not only in Louisiana where one might most expect them, but also in California's Central Valley, where they've dug miles and miles of tunnels in levees, seriously compromising public safety. Multimillion-dollar federally financed efforts try to eliminate the burly furbearer, whose contribution to the economy in the extremely limited market for fake mink coats was clearly not worth it.

Turning to the bigger picture, levees, like flood-control dams, have been designed for floods of the past. After notorious levee failures during the 1862 deluge, California's sprawling Central Valley has been converted to a long, skinny Holland by 1,600 miles of ever-larger-and-stronger levees, but they continue to fail with devastating results, notably in 1867, 1875, 1881, 1892, 1955, 1964, 1986, 1997, 2006, and 2023, and those are just the big ones. The state's pioneering water planner, "Ham" Hall, recognized the folly of looking only backward for future expectations and astutely predicted a plague of overtopped levees. He warned that a larger storm will always be coming; it's just a matter of time.[37]

The Army Corps spent hundreds of millions of dollars fixing California's larger levees after floods in 1986, only to have thirty major levees again fail and spill onto 18,000 acres eleven years later, prompting Jason Fanselau of the agency to comment, "We've never had levee problems spread this far and wide."[38] The high water of 1997 topped twenty levees in the small Cosumnes basin alone. Failures elsewhere allowed floods to close the West Coast's arterial Interstate 5 for a month and forced 24,000 people to flee their homes. As frequent as presidential elections, levee breaks have handicapped the Sacramento basin one out of four years on average since 1900.[39]

Explaining increasing damage, historian Robert Kelley pointed out, "Because of immense population growth, each overflow or levee break, when they do occur, is fantastically more costly and dangerous than in older times.... The Sacramento Flood Control Project was conceived and designed to protect farmers, and now it is having to protect large urbanized metropolitan areas holding populations running to the hundreds of thousands."[40] Echoing Kelley's grim retrospective of 1989, the number of failures grew by another 32 percent as of 2022, and the at-risk population increased from Kelley's hundreds of thousands to millions in the Central Valley today.

Extensive reinforcements had been made in the city of Sacramento, but outside it, multiple levees upstream broke along the Sacramento River again in 2006 and also at the river's terminus in the Sacramento–San Joaquin Delta, where farming of dried-up peat soil, combined with wind erosion, has resulted in a bizarre lowering of farmed and occupied islands to 30 feet below sea level—an elevation one might not expect to find outside the aptly named Death Valley. Tenuous levees have been built atop subsurface strata that is, in turns, both powderlike and jellylike—anything but solid. These precarious levees now guard half a million Delta homes that have usurped former farmland, which had been risky enough when covered only by crops. Now the residents occupy holes without drainplugs, all beneath the level of the sea, and their confrontation with gravity and oceanic forces relies entirely on pumps to lift accumulating water out. Those pumps will not work, of course, when power sources fail. Meanwhile, and unlike in the significantly safer low-spot of Death Valley, estuarine salt water, lapping directly on the outside of the levees, is projected to rise 2.5 feet by 2100, and many of the well-informed among us fear that the rise will be much more. Even though taxpayers spent $200 million on the Delta's unique problem between 2012 and 2022, the table for disaster has been set.[41] And just by the way, a massive earthquake is also due any time now in this quintessentially unstable terrain lying below sea level and showing every indication of becoming a marine environment just one tremor away. How can anyone think that this version of civilization will

The flood of 2019 ruptured or overtopped hundreds of levees across the Mississippi River basin, including here at the Platte River in Nebraska. Photo credit: Army Corps of Engineers.

work? How much will we collectively be willing to pay to prove that it won't?

With even higher stakes, the Mississippi is the nation's proclaimed "Father of Waters" but could just as well be nicknamed the "King of Levee Failures." The flood of 1927 caused 145 breaches that displaced 700,000 people from lowlands now housing far more. After various agencies spent sixty-six years addressing vulnerabilities at enormous cost, the 1993 Great Flood of the Mississippi caused 1,100 of 1,576 levees to fail—70 percent of the total. That's poor odds for all who find themselves living behind such tenuous fortifications. Worse, many far smaller private agricultural levees in the Mississippi basin are not immune to far lesser floods.[42]

Unlike the overflowing Mississippi that historically built fertile farmable soils whenever slack water oozed out across lowlands and quietly deposited a new blanket of fresh brown silt, the failed levees with their concentrated spills cause a different kind of flood. Blizzard-like, flows at failed levees variously erode soil and also deposit suffocating layers of coarse sand, both effects ruining crops and undermining the potential to grow more of them for years to come.[43]

In March 2019, with nine months of flooding only beginning to unravel, the Army Corps of Engineers reported that sixty-two Mississippi basin levees had already been breached or overtopped. Hundreds of miles of earthworks sustained serious damage, followed by floods that continued until autumn.[44] Exasperated Tom Bullock, commissioner of Holt County, Missouri, reported, "Breaches everywhere, multiple, multiple breaches."[45] Such devolution of the intended order reminds me of Mark Twain's unconstrained wisdom: "Ten thousand River Commissions, with all the mines of the world at their back, cannot tame that lawless stream, cannot curb it or confine it, cannot say to it, Go here, or Go there, and make it obey; cannot save a shore that it has sentenced; cannot bar its path with an obstruction which it cannot tear down, dance over, and laugh at."[46]

If a levee or floodwall has not failed, it has come hair-raising close at one time or another. Consider the big one. In 1993, St. Louis' "impregnable wall," rising 52 feet above normal flows, came within 2.4 feet of being overtopped, meaning that a 49.6-foot wall of water—five stories high and a third of a mile wide—raged on the other side of the city's razor-thin vertical defense of cement and rebar. St. Louis would surely have been flooded to multistory depths if levees directly upstream had not failed and released floodwaters across farmland that took the hit in self-sacrifice for the neighboring urban good—though none of those farmers chose to select this option.[47]

A prestigious National Academy of Sciences report remarkably concluded that one-third of all flood disasters in the US were caused by levee failures and cautioned, "It is short-sighted and foolish to regard even the most reliable levee system as fail-safe."[48]

Like dam failures, levee ruptures allow little or no time for evacuation, making this type of flood frighteningly more dangerous than that of a large undammed and levee-free river that encroaches inch by inch while storms and flows progress with comparative predictability and warning.

As part of a nationwide evaluation of infrastructure in 2021, the American Society of Civil Engineers graded America's levees with an unflattering D, though the official narrative reads more like an F.[49] The engineers projected that $80 billion was needed for repairs in the following decade, but nowhere near that kind of money was spent, appropriated, or even realistically imagined by the combined levee stewards needing the funds.[50] The Army Corps' levees weren't as bad, according to Corps reports, but still bad: 13 percent scored very high, high, or moderate with risks requiring "management solutions."[51] Even then, "management" solutions are no guarantee of solutions. As Professor Pinter grimly predicted for the long-term future of many rivers, "When the next flood comes along, bigger levees either fail or are overtopped again."[52]

LEVEES MAY BE UNDEPENDABLE, BUT THEY ARE NOT CHEAP

Most Army Corps' levees would never have been built without passing a benefit-cost analysis and without local sponsorship, which now requires commitment to a local or state cost-share and also to perpetual local maintenance, with "perpetual" being the heavy commitment of forever. Bearing weight on this multigenerational, nonvoidable marriage between the Corps and local government, and uncounted in original benefit-cost calculations, the costs of repair, upgrading, failure, and rebuilding add up, and such contingencies often exceed if not dwarf costs of initial construction. Sacramento's levees, for example, had already cost a lot through a century of upgrades but required $4 billion more after the 1986 flood.

Dwarfing other repairs, at New Orleans the Army Corps in 2019 reported that post-Katrina levee reconstruction costs topped $14 billion. But within another year—hardly time for grass to sprout on top—the new levee needed upgrading because repaired sections were already sinking into mucky substrate owing to the weight of the levee itself.[53] The substrate is what it is, and one might wonder how this problem will ever end for those who remain in the city; also for those who pay for the city's uncertain safety, and the circles of those two groups are far from coincident. The Corps reported that without further spending, subsidence combined with sea-level rise would cause aggravated "risk of life and property."[54] Ultimately the question may become, what can people afford?

Which takes us not so much to New Orleans with all its established neighborhoods and compelling charms, but most critically to rural areas where the economy of levee protection is abysmally worse than it is near any urban nexus, because in rural areas we face the undeniable fact: far fewer people depend on far more miles of defenses. Remember, 80–90 percent of nationwide levee mileage lies in farm and lightly populated areas. But that doesn't mean that levee construction there is easy or cheap. In California a new or rebuilt levee protecting a 200-year-frequency agricultural floodplain with few people living on it can cost $25 million per mile. That mile might only protect a few farms, and one has to wonder about the public cost of simply buying the flood-prone land and letting it flood—which has its own advantages of many kinds—if such a notion were not politically taboo for local politicians. But they tend to call the shots in these situations, and so the subsidies continue, in some cases racking up multimillions per farm, which might grow crops as profitable to the farmer as almonds but, then again, might only grow alfalfa or, in other words, grass.

The bottom line here is that while city people will be hard-pressed to pay their share for the levees they need, rural people will not be able to pay for their share at all. So who will do it? Anyone knowingly stepping up to this plate will want to know if the cost is worth the return and if the levees are sustainable or not. The likely outcome of this economic challenge is that rural levees will continue to fail. The floods will grow worse, the costs greater. In California, legal idiosyncrasies affecting many levees mean that the state is liable for damages when failures occur—an interpretation that alarms legal and hydraulic experts as well as taxpayers, and one for which the liability dust may not settle for some time. Yet the state does not prevent building in the danger zone, and so the potential damages—and public costs—continue to grow.

Let me note here that, according to calculations by the Corps, both its dams and levees nationwide averted $162 billion in flood damage per year, 2001–20, which is substantial on any ledger.[55] Yet, owing to new development constantly being built behind the levees and below the dams, annual losses continue to rise. Winning battles against specific floods, as the Corps claims, is no guarantee of winning the war against flooding—an outcome increasingly in doubt given our refusal to stop the continuation of development in the path of predictable damage, not to mention the problems of greater storms targeting aging—and therefore lesser—infrastructure.

The vulnerable new development occurs largely because the Federal Emergency Management Agency, through the National Flood Insurance program, greenlights insurance for new home-building behind levees that meet the Agency's engineering criteria, leading to ever-increasing investments in zones that will be flooded if and when the levees fail or, as we've seen at booming Natomas, simply when local pumps run out of power.[56] All this describes the safe-development paradox: increasing construction or hazard creep behind the levees creates vulnerabilities that accumulate with each new house, leading California journalist Elliot Diringer to describe a treadmill of risk and damage: "A century and a half after the first crude levees were thrown up, more people are at risk of flooding in the Central Valley than ever before."[57]

Like some other sticky and deceptive problems in life, the levee-building cycle is alternately self-serving and self-destructive: levees are built to keep floodwaters out of lowlands, people develop in the shelter of the levee, then larger floods overtop or break the levee, leading to larger levees, and so forth in a profound exhibit of the law of unintended consequences. Accumulated damages are devastating to the victims. Taxpayers shoulder many of the costs but reap few of the benefits. How can this go on?

As with dams, the question arises, what good is protection from minor and frequent floods if the levee is destined to ultimately fail in a large flood that cancels out the hard-earned savings of the past? To ask this another way, would we put our money into a savings account that earned a few percent interest per year if we knew that the same bank was ultimately going to go broke and lose all the money we've invested?

The reflexive response of many, in times of floods and after, is to double down with higher levees. But, as flood hazard specialist Dennis Milenti wrote, when we do that, we're not preventing damage but merely postponing it.[58]

BETTER LEVEES

Many levees are here to stay, and to evade costly and ineffective cycles of repeated damage will require new approaches. One of these is to move the levees back from the rivers' edges and give water the space it needs. How much should levees be set back and the river's pathway increased? The short answer is "as much as possible." Working against this logic, human nature after a flood is to repair immediately, and that means leaving levees where they are. It's easier to sandbag a hole or bulldoze some rock into a breach than to relocate an entire levee farther back from the water. The first option might take a day; the second a decade. Fortunately, some communities are beginning to navigate the literal and symbolic "high road" in realigning their levees.

Successful plans have also focused protection on small areas of intensive development with "ring levees" rather than extended linear ones that span miles up- and downstream. Dr. Raphael Kazmann, author of *Modern Hydrology*, contends that ring levees (actually shaped more like a U) around clusters of development can in some cases protect built-up real estate and leave the river free to overflow elsewhere. In Minnesota, East Grand Forks carried out a dual strategy of buying out 507 flood-prone homes and adding a discrete ring levee guarding what remained.[59]

More typically long and linear, and principally protecting rural areas, most levee mileage runs up and down both sides of rivers. To strategically improve these systems and escape from flood-raising straitjackets, California's Central Valley Flood Protection Plan calls for widening the corridor between levees when possible. Along the Feather River below Oroville Dam, the Three Rivers Levee Improvement Authority in 2010 widened its corridor for 6 miles, giving the river room to dissipate energy, in the process restoring 1,500 acres of exquisite valley oaks that thrive on groundwater-nourished floodplains. John Cain of River Partners noted, "By setting back the levees, you create room for the river and you decrease wear-and-tear on the hydrologic system." In another success, a lower Sacramento River corridor between levees was widened 800 feet, incidentally benefitting salmon and waterfowl.[60]

As we've seen, the falsely promised shelter of levees, along with FEMA flood insurance policies that recognize floodplains on the dry side of levees as reliably safe, have long led people to settle in the hazard zones directly behind levees, and as a result their newly built homes make it far more costly to move those levees back. Carson Jeffres at the University of California Center for Watershed Sciences cautioned that to relocate levees, "you really have to find those places that are primarily farmland." Because the vast majority of levee mileage in America is, indeed, at farms along the

Mississippi, Sacramento, and other large rivers, places can be found for this approach, though it still costs a lot of money. Farmers are reluctant to yield productive acreage to a relocated levee, whether it be for an orange grove or just an annual cutting or two of hay. Regarding the cost, Jeffres added, "It's interesting that the funding of restoration is considered expensive, but then when disaster happens, all of a sudden we're ready to deliver big disaster payments. Relocating levees is really an investment in our infrastructure to try to minimize those costs down the road."[61]

Widening the space between levees is good, but one has to consider that it might be cheaper to buy all the floodland and get rid of the levee than it is to reduce the amount of flood-protected land by building a whole new and expensive levee farther back from the river. Remember, in California the cost of constructing new levees, including those that are relocated for a wider corridor, was $25 million per mile, likely more today.

Aiming for a higher degree of restoration, and understandably less committed to the status quo in development and agricultural production on the American landscape, the Stillaguamish Tribe in Washington bought 1,200 acres of farmland in order to reestablish spawning and rearing beds for salmon by removing levees in the Stillaguamish River estuary. The tribe plans to restore another 6,000 acres as land becomes available.[62]

Another breath of hope came at a Nature Conservancy preserve in California. Floods in 1997 breached a levee along the Cosumnes River. Considering the lack of houses there, the Conservancy elected to not repair the damage. Since then floodwaters have repeatedly blanketed a wide cottonwood plain, restoring fish and wildlife habitat while posing no dangers to anybody.[63]

PLANNING FOR FLOODS OF THE FUTURE

What can people expect of levees in the future? The answer may vary between St. Louis, with a population density of 4,800 per square mile, and a typical acre of the Mississippi Delta, where the population in some areas dips to zero across a thousand acres of soy beans.

By far the greatest mileage of levees intended to protect flood-prone land has been built in rural and farming areas, such as here along the Sacramento River near Colusa, California.

Substantial urban populations live with risks behind riverfront levees—30 percent, for example, in Omaha, a city of high flood risks that's ironically known for its insurance industry. Cities dependent on levees will find that costly investments in safety must be made. But considering that only 7 percent of America is floodplain, and recognizing the low population density on most of that land, our expectations for protection behind the 33,000 miles of substantial levees and an estimated 99,000 miles of lesser ones are clearly out of step with economic or hydrologic reality. With rising floods and aging infrastructure, levees will not adequately protect a lot of the rural land that now awaits behind them.

Limitations of levees have been known for decades. But that doesn't mean that old-style pork-barrel funding for new levees is a relic of the past. Ongoing construction at Fargo, North Dakota, will deliver one of the most expensive levees ever built. As Dr. Pinter remarked, "With all the hallmarks of an old-time Army Corps flood-control project, hundreds of millions of dollars are being spent with a benefit-cost-ratio below one-to-one, making thousands of acres of floodplain available for new building. The pro-development lobby still parades through the halls of Congress seeking a continuation of those kinds of flood-control public works."[64]

In a vicious feedback loop, local communities seek to develop more land in the shadows of the levees in order to increase tax income to pay for upgrading or simply maintaining the same levees. As one engineer described the circular process, "It's almost impossible to generate the local funds to raise that levee if you don't facilitate some sort of growth behind the levee. You need that economic activity to pay for the project."[65] Yet new development too often results in greater damages when levees fail.

In an interview with *Sacramento Bee* reporters, John Garamendi, then deputy secretary of the US Department of the Interior and former California insurance commissioner, recognized the uncertainties of levees and advocated for a "national policy" to discourage further reliance on them. "We can't move Sacramento, so we have to protect it the best we can. But with regard to continued development, we should stop it, period. Just stop it."[66]

The future of flooding will require a new and dependable approach that improves on the tentative security of levees and dams. As Eileen Shader of American Rivers said, "We simply can't afford to keep trying to build our way to safety, relying on major structures to prevent damage."

A sensible approach beyond both the Stillaguamish Tribe and The Nature Conservancy's Cosumnes experience would be for the federal government to redirect funds now going to repeatedly damaged levees and repeatedly flooded real estate and instead acquire flooded farm acreage and rural homes when the cost of maintenance, failures, or improvements is greater than that of a buy-out. Over time, the aggregate acreage in restored floodplain open space could become, in effect, the greatest national wildlife refuge in America. Over time it could become recognized as the best money ever spent on control of flood damage. Sensible approaches, of course, are not immune to legal challenges and can be fraught politically, but long-term solutions clearly need to escape the current downward spiral of increasing floods, increasing damage, increasing costs, increasing debt, and increasing dependence on taxpayers living nowhere near the floods for solutions that, in the end, will fail.

For much of American history we've relied on dams and levees to protect us from floods, and we've either lost sight of other ways to solve our flooding problems, or never thought of them in the first place. The levees we already have can be made to work more effectively than they do, but wouldn't it be better to not need them at all across much of America? Let's next consider why this possibility will become increasingly important and compelling as the floods of the future grow painfully worse than the floods of the past.

6 Higher Floods and the Endless Storm

EVERYTHING IS CHANGING

Intensified floods and the vexing prospects of a heating climate will present ongoing tragedies, new opportunities, and revolutionary changes in nature, land use, and community—all within a future that's unexpected and, in many respects, unwanted. The changing climate is, as author Naomi Klein wrote, changing everything, and with it, the storms of the past will be reenacted in greater storms and in troubling patterns that appear to have no end.

When we nervously stare at floodwaters roiling to the tops of levees or quietly inching up across farmlands, we're seeing a climate crisis with global temperatures stoked high and effects spanning from desiccations of drought to a supersaturated atmosphere delivering storms never before imagined.

Floods are getting worse, and we'd do well to understand how much higher the high water of tomorrow will be. But before we consider how inadequately we're projecting flood hazards of the future, it's important for context, humility, and resolve, to recognize how badly we've underestimated floods of the past.

Think about the atmospheric river of chapter 2 that swept across the West Coast eight months after Civil War cannons were fired at Fort Sumter, even though that 1862 flood has largely been ignored. Furthermore, hydrogeologic analysis shows that within the past 800 years four floods dwarfed even that storm.[1] Few Californians are even aware of the greatest deluge in the state's history, let alone of alarming prospects for larger floods. Therefore one might say that the lack of concern by the average citizen is understandable.

What's less understandable is that carefully analyzed predictions, carried out in 2010 by an expert team of US Geological Survey scientists and based on a storm of the 1862 magnitude, provoked anemic responses even among people engaged in flood issues. In her book about natural disasters, Dr. Lucy Jones of the USGS and later the California Institute of Technology reflected on reactions to their forecast of an "ARkStorm," meaning one dropping 10 feet or more of rain in a relatively short period of time. "Many flood managers dismissed out of hand the possibility of that much damage. They knew what a flood was; they had managed many floods in the past. They wanted to believe that their engineering solutions could not be exceeded, so they ignored our findings." In a separate vein, Pacific Institute director Peter Gleick warned, "California is still not ready for another ARkStorm."[2]

Jones explained what she confronted as a "confirmation bias." Even knowledgeable people have rejected data that "didn't match their point of view."[3] This elephant-trap of denial, reaching extremes in the current political climate where dismissal of science and facts has become normal in some circles, can impinge on personal decisions, municipal management, and everyone's options for the future. Even the weather has become political, with large numbers of people eager to disbelieve what's offered by scientists, however reputable, when it conflicts with a preconceived view, however unfounded, outdated, disproven, or sourced in subterfuge aimed—even transparently—at boosting profits or political positions. Yet through this murky haze most people see that the weather is in fact getting more extreme, the heat hotter, storms wetter, floods higher. And who, regardless of political persuasions or doubts about cause-and-effect, would not want to avoid the hazards, the losses, and the pains that greater flooding will bring?

MAPPING A WORLD IN CHANGE

A lot of planning for disasters relies on projection of the "100-year flood" and the extent of land it covers. One might think that such a flood would reliably occur only once per century. But that's not quite right. This might sound like splitting rhetorical hairs, but the term means that the current year's weather has a 1 percent chance of scoring a 100-year flood, even if an identical flood happened last year. Still, factored over the very long term, the *odds* are that only one flood would occur per century to the limits of the 100-year floodplain. All that, however, depends on the mapping being correct, which is a high bar and, to mix metaphors here, a rapidly moving target.

Doing the math shows that, even when accurately mapped, a 100-year flood has a 26 percent chance of happening over the life of a thirty-year mortgage—poor odds when gambling with everything one owns being inside that flood-prone house, and when the house, itself, represents an entire life's savings.

Cancelling out those nest eggs of many people, the Lumber River in 2016 rose as a bloated nightmare with Hurricane Matthew's rainfall and all the flotsam it flushed into the stream's channel through rural North Carolina and the city of Lumberton. Hydrologic analysis revealed that a flood of similar height had only a 1-in-500 chance of occurring in any given year. To wrap our heads around this number that extends far beyond human lifespans, the best long-term expectation for seeing another flood of that level is the year 2516—a date more appropriate for really weird science fiction than for serious discussions of hydrology. At 1-to-500 odds for the whole year, none of us should expect to see a 500-year flood. Not even close.

Big surprise: two years after Matthew's drenching, Hurricane Florence delivered a "1,000-year" flood in the same place. Seventy percent of Lumberton—barely wrung-out—was submerged again in a sequence that was unexpected for centuries. But there it was in a two-year, wet-shoe turnaround.[4]

Poor Lumberton represents not only a glitch in data analysis, but also a troubling pattern. We've underestimated floods of the past and thereby underpredicted floods of the present, all of which leads to denial about floods of the future.

Lumberton's not alone. Milwaukee suffered back-to-back "100-year" storms in 1997 and 1998. And who cannot be impressed by Ellicott City, Maryland, where on May 27, 2018, a torrential rainstorm dumped 8 inches in two hours, turning Main Street into a Colorado River of brown suds, only this version was contained not by the Grand Canyon's golden cliffs of Navajo Sandstone, but rather by historic brick buildings awash in unmoored, semifloating cars smashing into walls, doors, windows. Amazingly, the scene reenacted one that had occurred only two years before. Also five years before that. Also twenty-seven other times back to 1768. The 2016 and 2018 events were each rated as a once-in-1,000-year event. Go figure.

Something's wrong here, and it counts, because any preparation for floods of the future begins with predicting the extent of land that's likely to be inundated. Though we're swamped in computer-age information, step-one in flood preparation has become enclouded by mystery over what "normal" actually is and by what "predictability" might actually be.

In fairness to the mappers, many projections are based on less than 50 years of rain and streamflow data, and rarely on more than 100 years of records—a blink of the eyes when dealing with timeless forces of nature. Stream gauges, beyond the familiar stick shoved into the sand—the level of technology that I engage in every night when I'm riverside camping on canoe or raft trips—were scarcely invented before the late nineteenth century, and some of these hard-earned monitoring stations have sadly disappeared owing to cuts in the US Geological Survey's budget. Streamflow data is based on rainfall data, which isn't much better. The National Oceanic and Atmospheric Administration (NOAA) updates rainfall records when it can with its "Atlas 14" reports, but much of that information, likewise, is fifty years old or more.[5] During Hurricane Harvey, some Texas stormwater agencies were using 1961 data for their warnings, like drawing on intel from the Eisenhower administration for foreign policy or even battlefield strategy during the Ukraine war. The mapmakers, of course, are using the only resources they have, but tenuous statistics lead to tenuous predictions, or to put it bluntly, the familiar "garbage in, garbage out" modus sets us up for a lot of error.

Innocently short on better data but inexcusably short on precaution, we use the information at hand to determine flood risks on which people's lives depend, not to mention the design of roads, bridges, sewers, dams, and levees gauged to last generations and promising severe consequences of failure if they don't. Storms come in single momentous onslaughts, multiple surges, and cycles that span centuries and millennia. But engineers' statistical sample is too short to accurately predict the next flood season, let alone an uncharted future spanning whole lifetimes with unforeseeable gyrations, local to global. To update rainfall projections, the Association of State Floodplain Managers persuaded Congress to add funding to the 2021 Infrastructure Investment and Jobs Act, authorizing new "Atlas 15" reports, a good start toward better data, but just a start.

Flood estimates can be overblown. For example, a volunteer geologist in Stonington, Maine, reviewed his town's flood boundaries and concluded that FEMA had overshot them, strapping residents with higher insurance premiums. The town successfully appealed to change the maps. However, a far more common and consequential error is that the mapping *underes*timates flood hazards, leading to dangers that should have been avoided. Soon after many 100-year flood maps are published, their findings are exceeded, which doesn't instill much confidence in the maps.

While estimates are based on historic flows and climate archives, spotty as they are, data assumptions also leave a lot of room for play. Where local streamflow records don't exist, mapping depends on hypotheticals based on nearby flood data. Worse, flood maps can become politicized weapons that variously protect, benefit, or cost money to what might generously be called a colorful cast of characters whose biases vary depending on their station in life and role in the community, not to mention entrepreneurial ambitions and human slippage in ethics. Wherever big money is involved, even tightly prescribed data can be spun, especially in small communities without adequate professional staff to check the numbers. The bottom line here, as W. Craig Fugate, FEMA administrator under President Obama reflected, is that "local governments have been opposed to any maps that show an increasing risk."[6]

Take Livingston, Montana. Recognizing inadequacies of a local levee, the Army Corps of Engineers mapped the floodplain accordingly. Local developers complained, the city hired a private contractor to redo the flood maps, the state's senators interceded on the city's behalf, and FEMA in 2011 agreed to a 95 percent reduction of the floodplain, only to see major flood damage in the disputed territory in 2022, when the same floods closed access to Yellowstone National Park nearby. Political scientist Sarah Pralle of Syracuse University noted that such "politicization" of flood maps is common. An NBC News survey quickly found 500 instances where FEMA, presumably under requests—and those requests may have morphed into what might be called "pressure" if the Livingston case provides any clue—had remapped flood-prone properties into lower-risk categories.[7]

Flood predictions for some communities, Lumberton and Ellicott City being Exhibit A, here, were blown away repeatedly within a decade. FEMA's own postmortems humbly indicate that one-third of flood damage occurs outside the agency's officially mapped 100-year floodplains owing to underestimated runoff and to communities where floodplains have not been mapped at all.[8] This is true not only in rural areas presumably back-burnered from mapping efforts, owing to low population densities, but in some urban districts as well. Analysis by the First Street Foundation found that Chicago's Englewood neighborhood had some of the city's most-flood-prone land, one-third of properties at risk, though hazards were not evident on FEMA maps, at all. The area happened to be 95 percent Black, but overflows of small streams traditionally go uncounted by FEMA no matter who lives there.[9]

Another problem skewing forecasts and causing floods to be higher than predicted is that new and unaccounted-for development alters the size and timing of floods in municipalities without adequate stormwater management rules. This includes most communities. Substantial upsurges result because new impermeable pavement and roofs cause stormwater to run off immediately where it used to soak into the ground and filter back to streams slowly. FEMA reports that with ground cover of trees and grass, 15 percent of total rainfall runs off, but with urbanization, 61 percent flushes immediately.[10] The contrast can be even more dramatic. Pristine watersheds of deep forest cover can absorb significantly more water than FEMA reported. Large impervious surfaces, on the other hand, shed every drop; consider a new Walmart's sea of asphalt and continent of rooftop—a sure recipe for instant runoff unless innovative drainage systems are deployed. Even without such extremes, typical sprawl along Johnson

Creek in Portland, Oregon, caused flood frequencies to increase from once per 100 years to once in 25, all describing a substantial downside to suburban patterns of growth that are typically rich in sprawl, big-box stores, and pavement.[11]

Similar effects or worse result when wetlands are drained, marshes filled, and hills flattened for subdivisions, not to mention whole horizons clear-cut down to bare dirt by logging companies, causing runoff to come faster, peak higher.[12] In Appalachia hydrologists found that strip-mining for coal can increase flood runoff by a staggering 1,000 times.[13] All this adds up, but FEMA does not account for projected land-use changes, at all, in its flood-mapping efforts.

Droughts in many cases don't lessen the prospect of floods, as one might imagine, but perversely worsen them. In the global-warming era, desiccation, particularly in the West, leads to wildfire, which acutely aggravates floods that follow. A UCLA study reported that when more than one-fifth of forest acreage burns, peak streamflows increase by 30 percent for six years owing to lack of vegetative cover. Consider that outcome when applied to the combined western states, where acreage that annually burned grew by 1,100 percent between 1984 and 2020, with severely worse prospects in sight.[14] Climate scientist Danielle Touma of Stanford University projected "a future with substantially increased postfire hydrologic risks across much of the western United States."[15] Nor are dam or levee vulnerabilities considered when mapping floodplains, yet failures happen and cripple whole communities and regions.

This litany of error is discouraging if not overwhelming and perhaps paralyzing. But wait. Here's the most-important among all the reasons for higher-than-expected floods: the worldwide climate is getting hotter. This inevitably causes storms to grow larger. In 2022 FEMA Administrator Deanne Criswell recognized, "Climate change is overwhelming US flood maps."[16] Most FEMA maps do not reflect the rapidly escalating projections, yet global warming—to say it again—is changing everything. Previous floods occurred on what journalist Bill McKibben described as "essentially a different planet with a different atmospheric chemistry."[17] Today it's a whole new game.

FLOODS ON STEROIDS

According to NOAA, over a period spanning 119 years, 1895–2014, nine of the ten years of highest precipitation in eastern and midwestern US occurred after 1990.[18] Some Mississippi River residents have experienced a "500-year" flood every five to ten years since the megaflood of 1993.[19] For 2016 alone NOAA classified eight floods as 1-in-500-year events.[20] In 2020 a First Street Foundation study estimated that flood risks were 70 percent greater than what FEMA maps predicted.[21]

Anecdotal evidence of increasing floods has cropped up everywhere. On three separate occasions in 1997 California's San Joaquin River flooded higher than previous records.[22] At the Santa Cruz River in Tucson, Arizona, the 100-year flood was projected at 11,400 cubic feet per second—a lot of water for a typically bone-dry streambed. But in 1983 the river wreaked havoc at 52,700 cfs. That's enormous even for the regionally arterial Colorado River. Given this single new data point, engineers upped the 100-year event for Tucson by 500 percent. Nationwide, new data points are occurring all the time. Flood projections are going up and up and up.

In Houston, "500-year" floods occurred in 2015, 2016, and 2017. During Hurricane Harvey, alligators, not to mention brown water snakes, swam in neighborhoods that no one ever expected to flood under any circumstance. Whole city blocks became a reptile-land of the Gulf Coast. Half the insurance claims lay outside mapped floodplains, and in adjacent municipalities 80 percent of submerged land had not been considered floodable.[23]

Houston's off-the-charts storm made evident that no one really knows what the upper level of future climatic events will be, except that they'll be larger than anyone has imagined. As General Gerald Galloway, formerly of the Army Corps of Engineers, recognized, "Our expectations for flooding are based on statistics of the past, but unfortunately the future is going to be different."[24]

The underestimation of flooding has only gotten worse since 1997 when FEMA financial manager Ed Pasterick modestly admitted to the *Sacramento Bee*, "We've got a whole legal and financial system that depends on floodplain maps" and that "we put way too much reliance on where those lines are drawn."[25] With rhetoric unhobbled by FEMA's bureaucracy, Marshall Shepherd, director of Atmospheric Sciences at the University of Georgia, noted that the 100-year metric, as typically measured, "is pretty much useless now as a baseline for an extreme event."[26]

Here and in other cases as well, flood expectations have been vexed by a classic failure to respect the precautionary principle. This rule of survival, old as the jungle, urges that, when faced with uncertainty, one should be careful. Slant expectations toward the safe side. Caution over calamity is the path to longevity. Animals, individual people, tribes, communities, nations, and whole species that have embraced this axiom are the ones that have endured the rigors of time. Evolution has favored those who recognize that bad stuff happens. Surprising uncertainties include the entire planet heating up, which no one envisioned when many of today's flood maps were rolled out. The unexpected includes the whole world changing, which, with global warming, is now an undeniable, unchangeable fact of life.

As an efficient way to incorporate precaution, professor Sam Brody of Texas A & M University advocated building codes requiring that new homes be elevated not just above the 100-year-flood elevation, which FEMA often requires for subsidized flood insurance, but 3 feet above it—clearly a step in the right direction, but probably not enough.[27]

An important corollary to the precautionary principle is that its importance grows as risks increase. The principle also becomes more important as expectations become less certain. Both these axioms apply in spades to modern flooding, though that observation seems to be ignored, even while prospects of future high water push risk factors off the charts, and even while climate change causes sharp escalations of the unknown, altogether a "perfect storm" of miscalculation riddled with a lot of willful ignorance among many people, including some making pivotal decisions at local, state, and federal levels.

Much in our mapping, our regulation, our investment, and our human response to flooding appears to have rejected the precautionary principle altogether. Though FEMA delineates the 500-year floodplain when drawing Flood Insurance Rate Maps, that level of safety is rarely the regulatory baseline, meaning that it's essentially ignored. Municipal governments typically adopt the lowest plausible regulatory minimums, even while evidence piles up that the highest floods of the past will be exceeded. Or, one might say that the worst-case scenario is the one we should be expecting—if not a worse worst-case, as the climate news grows inexorably more dire.

Our primary approaches to preventing flood damage—dams and levees—have, at best, offered partial solutions, creating what has turned out to be a losing game: one-step-forward with infrastructure, two-steps-back with damage. Now, in addition to facing historic inadequacies and our chronic failure to address sources of flooding problems, a new era of previously unimagined flooding is upon us.

IT'S THE CLIMATE, EVERYONE

"We've never seen anything like this before." Many people repeat this refrain when facing flood hazards of recent years, and also as wildfires rage through western forests, rangelands, and suburbs, as people literally drop in heat waves that smash record after record, and as glacial ice of the centuries drips away to sea in volumes equal to some of the Great Lakes. In 2021 the US endured twenty separate billion-dollar climate disasters, many involving floods, together making that the third-costliest weather year on record. The numbers of extreme events through the previous five years averaged 17.2 compared to 5.3 in the 1990s.[28] All but the most jaded doubters and science-deniers agree that intensifying storms, increasing amounts of rain falling in short amounts of time, and flashy torrents of runoff are attributable to the changing climate.

No one at this point in the decades-long debate about global warming should be fooled by the difference between weather and climate. Weather is what happens now. One particular snowstorm on a wintry day in February 2015 prompted James Inhofe, no less than chairman of the Senate Environment and Public Works Committee, to carry a melting snowball onto the floor of Congress as his version of proof that global warming was a "hoax." The senator provided evidence only that it had snowed that day in Washington, DC. That year, actually, ranked globally as the warmest on record, a climatic fact that seemed to be of no interest to the oil-state senator.

The problem starts with our burning of fossil fuels, our cutting and disposal of mature carbon-rich forests, and our depletion of soil-based carbon by exploitive farming techniques. All these activities have dramatically increased the amount of carbon dioxide in the atmosphere. The big culprits are fossil fuels—coal, oil, and methane gas—whose burning produces carbon dioxide, which lingers airborne for centuries and acts like the roof of a greenhouse, allowing the sun's heat to enter but perversely preventing its escape. Methane from uncapped and leaking gas wells is far worse than carbon per unit emitted, but that's another story.

In 2021 the amount of atmospheric carbon dioxide reached 419 parts per million, highest in 4 million years and increasing by 40 billion metric tons annually according to NOAA. The last time carbon registered so high, sea level lapped 78 feet above today's shorelines. The rise in sea level lags behind the rise in carbon, which might cause one to question how high the oceans will encroach again.[29]

Over half the human-caused carbon dioxide produced since industrialization has been generated since 1988. No coincidence, but rather in direct relationship, and purging any thought that a year or two were just freaks, the world's eight hottest years occurred between 2014 and 2023.[30] The hottest was 2016, and 2021 ranked as sixth-hottest with broiling temperatures nationwide and mind-boggling heat in the famously cool Pacific Northwest.[31] July 2021 was the hottest month known.[32] March 2023 was the second-hottest March. New records will likely be set before this book is published.

For years the climatic shift was descriptively called "global warming." That term was traded out for *climate change*, a phrase regarded as more inclusive by covering precipitation patterns gone strange and quirks of meteorology such as hurricanes. But *climate change* also neutered the image of Earth getting hotter, which was not a message the fossil fuel industry liked. Global warming was renamed without a whimper of dissent, and the Earth's fate was reinvented as something we all know to be the normal state of affairs: change. That's accurate enough, but if anyone had wanted to soften the impact of climatologists' warnings without actually confronting the data, *climate change* would be the take-home phrase. Recognizing that the warming of Earth's atmosphere is the root cause of virtually all the relevant climate changes we're seeing, I prefer the old term, if not *global heating*.

Grim as the projections are for global heating, they've often been exceeded in recent years, and many by the proverbial mile. Carbon emissions—the root problem—have risen not just above predictions of experts but dramatically above worst-case scenarios earlier forecast.[33] University of Miami geology professor Harold Wanless explained that the United Nations' Intergovernmental Panel on Climate Change—the gold standard for forecasts—has chronically underestimated the problem. IPCC forecasts from 1990, 1995, 2001, 2007, 2014, 2022, and 2023 grew progressively more dire. Wanless maintained that the IPCC uses data published in referenced journal articles, which take time to write, and then the findings undergo peer review, which takes more time, and together the delays can consume up to a decade for a problem needing attention daily. Scientific consensus is virtuous, but Wanless warned that when all is done, we've dangerously underestimated the bind we're in.[34] Of course others, with a far greater grip on mass media than either Wanless or the IPCC can claim, have been eager to underestimate the global warming problem, if not to deny its existence.

The facts get scarier all the time. A 2022 study published in *Nature Reviews Earth & Environment* reported that 90 percent of the heat caused by greenhouse gases is eventually taken up by the seas. The rate of warming there has doubled since the 1960s and will double again by century's end. Beyond the decline of ocean habitats and loss of life, ranging from polychrome corals to blue whales, this warming will fuel hurricanes and rainstorms like never before seen.[35] In spring 2023 NOAA reported unprecedented ocean temperatures, extremely troubling to oceanographers and climate scientists.[36]

Melting ice and ocean warming have caused sea level to rise 8 inches since 1900, 3 inches since 1993 alone—fastest rate in 2,000 years.[37] NOAA projected a national average rise of 10–12 more inches by 2050, with levels along the western Gulf Coast increasing a foot and a half owing to localized ocean currents, tides, and subsidence.[38] Far greater encroachment will come with melting ice caps in Greenland and Antarctica, forecast to mount sharply after 2100 but a wild card in any projection and one that could be played by the gods of climate at any time.[39] Consider 2021; for the first time in history, rain fell on Greenland's glaciers at elevation 10,500 feet and accelerated melting on an ice mass three times the size of Texas, which, just for perspective, takes sixteen hours to drive across. In the past decade-and-a-half, net glacial runoff worldwide has poured the equivalent of Lake Michigan into the oceans. Lake Michigan is over 300 miles long. A lot of additional ice—solid water now beneficially stacked up above sea level—will soon melt and raise the oceans dramatically. A 2022 study projected that the seas will rise at least 10 inches from Greenland melt-off alone by century's end.[40] Even greater encroachment of oceans owes to the fact that warmer water occupies more physical space than did the previous cooler liquid of greater density. This distant precept from high school physics class has major real-world implications when applied to the combined oceans covering 70 percent of the globe.

Because of flooding from rising seas, millions of people worldwide will need to move. Additional millions will flee flooding rivers; consider the 7.2 million affected by Bangladesh floods in July 2022. In August floods in Pakistan displaced 33 million people temporarily or permanently.[41] Climate refugees of the future will put the already intractable and politically divisive refugee migration crises of the 2020s to shame. No one knows how this redistribution of world population will roll out. Along whatever painful path one might imagine, scenarios that we'd prefer to not even consider will become inevitable.

If we halted all fossil fuel burning tomorrow, effects would still last centuries. If we don't halt widespread use of fossil fuels, the effects could make much of the Earth uninhabitable.[42]

So, what to do? First, reduce the sources of atmospheric carbon and methane pollution—principally our burning of fossil fuels. Efforts to do this are familiar to all and may constitute the most important mission of humanity. Second, minimize effects of the changes. This process of dealing with the unwelcome inevitable is now called "mitigation," and it includes addressing the global-warming floods of the future.

Faced with not just the science and evidence but also the climate catastrophes of the month, week, and day, the American public has shifted toward a tipping point. One reliable source indicated that public opinion in 2022 had grown to 61 percent of Americans thinking that Congress should do more about global warming.[43] While society grapples with underlying causes, we'll have little choice but to address global heating's effects already underway, including intensified rainstorms.

INCREASED WARMING CAUSES GREATER STORMS

For every 1 degree F temperature increase, the atmosphere holds 4 percent more water, all of which eventually falls to earth. No surprise: virtually all climate models project that future storms will deliver more rainfall than ever before.[44] Because of global warming, Hurricane Harvey's rainfall in Houston was estimated to be 40 percent greater than it would otherwise have been.[45]

NOAA data in 2022 showed that many rainstorms already deliver 30 percent more water than those of the recent past. What has long been considered a 100-year storm now occurs once in 30 years.[46] That's just the beginning of a troubling trend. Take your pick of reputable projections here. The *Fourth National Climate Assessment* by the US Global Change Research Program in 2018 forecast that precipitation will increase up to another 40 percent across much of the country.[47] The National Center for Atmospheric Research warned that the number of extreme rain events could increase 400 percent by 2100, delivering 70 percent more water nationwide.[48] The Department of Energy's Pacific Northwest National Laboratory reported that total precipitation there will likely swell by 30 to 40 percent by midcentury under a high-warming scenario.[49]

In 2022 Ryan Harp of Northwestern University reported that actual precipitation records confirm what the models have been projecting: more water is being dropped per storm across nearly all the US, especially in eastern and central states.[50] Few people would ever have thought that the Spanish term *derecho,* meaning a supercharged, fast-moving complex of thunderstorms or "inland hurricane," would enter the weather watcher's lexicon in places such as Missouri. But it has. The United Nations' Intergovernmental Panel on Climate Change issued its sixth climate report in 2021 cautioning that "heavy precipitation" will fall with "extreme events unprecedented in the observational record."[51]

The only areas in the US without warnings of greater annual rainfall were the southern Great Plains and Southwest, both desiccated by droughts of the global-warming era. But, as if that region's gasping thirst were not curse enough, the intensity of summer monsoons will also grow, as they did in 2021 with downpours triggering flash floods. After the rain ceased, the region was promptly parched with drought again.

In California, Dr. Daniel Swain at the National Center for Atmospheric Research reported that owing in part to global warming, the chance of seeing a mega-storm on the scale that swamped the Central Valley in 1862 is "almost an inevitability" by late in the century.[52] Swain and climatologist Xingying Huang forecast that an atmospheric river of centennial proportions will likely displace 10 million Californians and cost $1 trillion in the largest "natural" disaster in history. Runoff from the Sierra Nevada will likely be 200–400 percent more than in past storms, leading not only to the highest floods ever recorded but to landslides, debris flows, and perils of weakened dams and levees.[53]

In 1972 Hurricane Agnes delivered the most destructive storm-and-flooding event in American history, but in the half-century since, fifteen storms each caused more damage than did Agnes, some so bad that the cynical among us might be tempted to call the 1972 disaster "quaint" with its hit of $2.9 billion versus $169 billion for Katrina and $130 billion for Harvey, all in 2019 dollars.[54]

According to Four Twenty Seven, a business firm estimating financial market risks and part of Moody's Investors Service, rainfall increases are coming and will most heavily impact the central Appalachian Mountains of West Virginia along with Ohio and Kentucky, followed by the rest of the East and South.[55] Sure enough, 17 inches of rain broke Tennessee's daily record by 3 inches in 2021; double the increase meteorologists had considered a worst-case scenario.[56]

Figuratively speaking, the weather now is all over the map. Global warming will exacerbate already-evident patterns of droughts followed by floods.[57] This dry-wet knockout is familiar from days predating our knowledge of global warming: floods of 1936 followed apocalyptic aridity that sent Oklahoma dust storms east to grit US Senate chambers even with the windows shut. Severe droughts in California were followed by floods in 1955 and 1964. With global warming picking up the pace of the newly termed "weather whiplash," a 2021 drought ranked as the century's worst with western wildfires smoking New York City 2,500 miles away, all upended overnight in October with record-breaking rainfall when an atmospheric river hosed California and flooded streets, San Francisco to Los Angeles. July 2021 was California's driest month on record, and December saw one of the largest amounts of snow ever, followed by history's driest January and February. Meanwhile flash floods ravaged the Southwest and East. Citrus grower Dale Murden of southern Texas lamented, "I've lived through hurricanes, freezes, and droughts, but never in my life have I experienced all three in one year."[58]

Global warming was credited for unprecedented flash flooding in 2022 with storms that killed thirty-seven people in Kentucky, hydrobombed St. Louis, and drenched our most arid city, Las Vegas, with boatable water riffling down casino-lined streets of the Nevada desert. "The extremes," former California governor Jerry Brown informed us, "are the new *ab*normal."[59] California's driest three-year period in history occurred in 2019–22, abruptly ending in winter 2023, with record-breaking rain and snow delivered by not one atmospheric river but seventeen to thirty of them, depending on how they were counted.[60] Within a few weeks Californians awkwardly pivoted from the state's most severe drought to one of the deepest snowpacks, quickly infused by penetrating depths of rain leading to broken levees, overfilled dams, and flooding. Tulare Lake—once the West's largest in acreage—had been dried up by agribusiness diversions for decades, but floodwaters of 2023 refilled much of it.

GREATER STORMS CAUSE GREATER FLOODS

We've already seen that floods are increasing. Now it's evident that increases are aggravated, if not caused, by global warming, and therefore the floods grow larger, no end in sight. At the central and south Atlantic states, coastal flooding swelled by 125 percent, 2000 to 2015, and the walls of rainfall encroached far inland.[61]

The wind-driven speeds of storms' progressions are slowing down. That might sound like a moderating force, but it's not. When a storm stalls over a landmass, it continues to rain at the same place, producing greater floods than what occurred when clouds formerly paraded by on prevailing winds. NOAA's James Kossin reported that storm migrations slowed 10 percent from 1949 to 2016, concluding, "Nothing good comes out of a slowing storm" and that the altered cycle means "more freshwater flooding."[62] This phenomenon occurred in spades as Hurricane Ian, in agonizing slow motion, drenched Florida with record-breaking cost in 2022.

Meanwhile across the West, for each 1 degree F increase in atmospheric temperature, average snow lines retreat 300 vertical feet up mountain slopes. An expected 5.4-degree warming by 2100 will push snow lines 1,600 feet higher, covering a lot of acreage when spread across the 400-mile-long Sierra Nevada uplift. Even if total precipitation remains the same, we'll get more rain and less snow, which means a lot more water coming at once, or, in other words, a flood. The Sierra provides up to 80 percent of California's drinking and irrigation supply, and its natural pace of delivery has for ages been rationed over a long spring and summer totaling five months of melting. But rising temperatures mean that those mountains will lose 40–80 percent of their accustomed snowpack by 2100.[63] More rain, combined with the rain-on-snow phenomenon, could triple current sieges of runoff.[64]

Crunching these numbers in economic terms, scientists at the University of Bristol and University of Pennsylvania analyzed flood records, insurance claims, and census projections, then warned that flood damages in the US will increase by 26 percent in the next three decades owing to global warming alone, provided we meet carbon-reduction targets. If we don't—and thus far we've not come close—flooding will be worse. In a separate approach, the First Street Foundation reported that the warming climate will cause flood losses in US homes to rise from $20 billion in 2021 to $32 billion in constant dollars by 2051.[65] That forecast was based not just on floods of the past, like the forecasts by FEMA, but also on median global warming predictions, new development anticipated in hazard areas, sea-level rise, and new detail in mapping.[66]

Professor Daniel Swain and his team found that increased rainfall will cause an "essentially inevitable" 20 percent surge in the volume of floods and a 200 percent jump in their frequency, leading to a 30–127 percent increase in numbers of people at risk. Swain warned, "Such a reality points to the urgent need for targeted policy and land use interventions to guide future development away from these high risk zones."[67]

The not-so-hidden script is that flood maps need to be redrawn and risks described more accurately.[68] Long criticized for not updating projections, FEMA in 2013 began to consider climate change and warned that the 100-year floodplains of the past are likely to see a 45 percent increase in area nationwide.[69] Even if one doubts that floods are increasing, or

Watershed disturbance caused by urban pavement, farming, and clear-cutting of forests all heighten modern flood levels, but the greatest increases now come from intensified storms of the global-warming era, which occurred here at Guerneville, California, in January 2023.

even if one questions that the source of higher flows is global warming, the increase in development that's prone to flood damage is profound.

As scientists refine the analytics of climate, flooding, and damage, the news grows more dire. Using state-of-the-art climate models, First Street Foundation analysts reported that a quarter of the nation's roads, or 2 million miles, may become impassable during floods at one time or another. One-quarter of America's critical facilities, including hospitals, fire stations, airports, and municipal buildings, will suffer flood-induced shutdowns of some degree, and those are the last things anyone wants to shut down during an emergency.[70]

All this aside, some analysts question whether an increase in overall worldwide flooding has been statistically evident in the global warming era.[71] The shortness of historic streamflow records makes determination difficult to prove by the most-cautious standards, but even skeptical projections express little doubt that the increasing short-term intensity of rainfall is causing more extreme storms and greater flash flooding.[72]

Consider the nationwide trend: in 2016 five floods, each rated as a 1-in-1,000-year event, all occurred within one year, giving cause for alarm. Then in 2022 another five floods were also rated as 1-in-1,000-year events, but this time all those floods occurred within *five weeks*.[73]

Especially threatened by coastal flooding, and also by swollen low-country rivers in the path of hurricanes, are Florida, Texas, North Carolina, Louisiana, and South Carolina—all "red" states whose politicians typically deny realities of the climate crisis. Intensified river flooding will be most extreme in West Virginia with its cramped valleys hammered by infamous rainstorms. Not only are many of Appalachia's homes tucked into the bottoms of those gaps pinched between rugged mountains—ground zero for flooding—but 61 percent of the state's powerplants are vulnerable to inundation, with the irony that they burn the worst source of global warming: coal. Having so much to lose—actually, having so much already lost—West Virginia might be expected to lead the charge against global warming, but logic matters little where Senator Joe Manchin, a coal tycoon, single-handedly blocked and then weakened the Biden administration's most explicit efforts to fight climate change in 2022.[74] Addressing the universality of this problem, co-chair of the Intergovernmental Panel on Climate Change Professor Hans-Otto Pörtner said, "Political will, in terms of climate action, is the bottleneck for a sustainable future."[75]

TOWARD A NEW APPROACH

Informed people now know that climate change is real and that floods grow worse, yet motivation to address the problem lags colossally behind severities of threat. A substantial number of congressional members remain under the influence of climate deniers funded by the fossil fuel industry.[76] Denial is the easy path when a crisis is not in our face. But the floods are, with high water coming in greater force, suddenness, and frequency. Our response will indicate our ability to choose—or not—a path of resilience and survival for natural systems and communities alike.

The tendency to build houses and businesses on floodplains has gone from successive and overlapping eras of perceiving development there as an opportunity (cheap land), as an economic necessity (affordable real estate), as a personal desire (a view to the water), and as a tempting option where risks are softened by subsidized dams, levees, relief, and flood insurance. But storms of the future will cause all the lures of floodplain development to fade. In the contest between infrastructure investment and risk, the perils of flooding are going to win.

So what do we do about that? We've maxed out our options with dams and levees, building wherever flood-stopping structures made it seem sensible to build plus in a lot of places where they didn't. The response to the floods of tomorrow must extend beyond our approach in the past.

Journalist Brooke Jarvis perceptively questioned society's wisdom as we stubbornly "weigh known benefits against unknown costs and chose to move ahead."[77] Sure enough, with known benefits but with costs still obscure to many, and clinging desperately to denial, we continue to build homes where floods grow worse. Disbelieving or ignoring forecasts, we show little inclination to do what our survival requires. Yet the climate of the future consigns our society to failure in our development and occupation of floodplains.

The age of denial is over. The time has come to take a different path.

7 Floodplains Are for Floods

THE FLOODPLAIN MANAGEMENT ALTERNATIVE

Beyond building levees, erecting dams, and reimbursing flood victims with disaster aid, a fourth alternative to address flooding problems is quite simple: avoid the paths of floods. Just stay out of the way. Yet floodplain "management"—principally to keep floodplains undeveloped—has become complex, challenging, and vexed.

The avoidance option was surely intuitive ever since people first stood gazing out at a river and knowing essential truths. The water will rise. We are terrestrial creatures. But somehow the survival instinct was lost, or at any rate fell behind other imperatives.

To address any issue, and clearly that of flooding, rule number one might be to not let a situation get worse while we're trying to make it better. If the bathtub's overflowing, mop up the floor, but first turn off the spigot. In the context of this book, that means not building on floodplains that remain as open space. Rocket science is not required to know that stopping development in danger zones is the safest, fastest, most efficient, and least expensive alternative toward reducing future damage. It's also

Flooding of undeveloped land results in relatively little economic damage, even along farmland such as this bordering the Susquehanna River in Pennsylvania. About 90 percent of floodplain acreage in America remains, thus far, with comparatively little development and can be protected as open space through effective floodplain zoning.

the only path to protecting and restoring natural floodplain functions on which rivers, ecosystems, and whole economies depend.

The importance of keeping new investments off floodplains is paramount because undeveloped or sparsely built-up property still accounts for the lion's share of flood-prone land—an estimated 90 percent nationwide.[1] As bad as flood damages are, they could get much worse because we're not doing enough to protect that land from new development that would flood.

With the rapid ascendance of engineering in the twentieth century, along with its kindred technical and construction arms, this cautious "floodplain management" alternative was seemingly unconsidered in the hubris of short-term economics and boom-time building. As we saw in chapter 2, even our wisest political leaders through many years, including FDR and JFK, governed as though we could "stop" or "control" floods, but looking back to history and ahead to climate projections, it's clear that we can't. Admitting this is not defeatist, but simply recognizes that everything exists within the rule of natural law. To think we can eliminate floods is like thinking we can cancel gravity. Of course, we know we can't do that, so we don't go jumping off skyscrapers. Yet when the consequence of risk is not immediate, people respond to the inevitable differently. Some are inclined to fight nature; others to fit in with it.

Surely avoiding damage would not be simply an "alternative"—as in the "floodplain management alternative" of this subhead title—but rather the first option for reasonable people. Nobody thinks that to avoid breaking a leg is merely an "option" to emergency care and then wearing a cast, walking with crutches, and sweating through six months of physical therapy. Not breaking a leg is our first choice. No, it's not a choice at all, but rather an imperative needing no discussion. Better to avoid a clear and present danger than to regret underestimating it. Yet here we are, with "management" of where we construct buildings as the "alternative" to the dominant approach of costly dams, undependable levees, and endless needs for disaster relief.

Beyond the realm of common sense, economics and public expenditures call into question our obsession with dams and levees rather than soft-path approaches such as restraint when choosing where to build homes. Instead of taking personal responsibility for where we live, we've opted for the Army Corps of Engineers to spend $123 billion of public money (adjusted to 2011) on structural flood-control projects. And that's just the Corps. We've funded other mega-billions in subsidies, structures, and relief for floodplain development instead of floodplain protection.[2] All with disappointing results.

For a long time the bottom line has been bad, but now the National Oceanic and Atmospheric Administration reports that the costs of climate-disaster events including flooding in the US are rising sharply. In only the five years 2017–22, losses totaled one-third of all such costs for the recent forty-two-year period, adjusted to inflation. The years 2017, 2005, 2021, and 2022 ranked as the costliest on record.[3] Wrapping up the sad state of affairs, Laura Lightbody of Pew Charitable Trusts cautioned, "The damages continue to increase, and one reason is that we keep building stuff in the way of floods, and it's expensive stuff. Many local officials put a higher priority on short term economic development than on cutting risk in the long-term future."[4]

Not just a problem of the US, global failure to recognize increasing perils of new floodplain development will affect hundreds of millions of people. Worldwide, 86 million moved into flood-prone areas between 2000 and 2015. Projections for 2030 by geographer Beth Tellman of the University of Arizona call for an additional 179 million floodplain residents by 2030, presumably followed by more.[5] While paying a lot to reduce flood hazards, we're allowing the problem to grow ever more severe. Instead of providing a desperately needed model for the international community, here in the US we've failed to do what's needed to adequately protect floodplains from further development.

GENESIS OF CHANGE

Though it's an afterthought to a century of national policy seeking to "control" floods, insightful leaders, policy analysts, professional planners, and advocates for river conservation have in recent decades taken increasing strides toward a nonstructural, natural systems, and management approach to minimize flood damage, though the outcome is far from resolved.

Let's go back: to curb the increase in losses, the protection of floodplains as open space was championed by geographer Gilbert F. White as early as the 1930s. Motivated by the Quaker faith to aim his research and influence at topics beneficial to society, White grew fascinated by the problems of natural hazards. At the impressively young age of twenty-two he served in FDR's budget office and worked for the administration's Water Resources Committee. In 1938 White feared that the wrong message was conveyed in pending legislation that eliminated local cost-sharing for flood-control works, and he recommended a presidential veto, cautioning that if the bill passed, "all states and communities will forever think flooding is a federal problem for the federal government to solve."[6] FDR signed the act, reportedly with personal apologies to White.[7]

While in graduate school at the University of Chicago, White wrote his landmark dissertation "Human Adjustment to Floods," blazing his intellectual and professional path to the future. In the following decades he chaired many commissions on flooding and disasters, founded the Natural Hazards Center at the University of Colorado in 1974, then directed that clearinghouse and research organization aimed at making floodplain management information available to the public, emergency managers, and agencies.

Early on, White considered the rapidly accumulating costs of floods and famously concluded, "Floods are 'acts of God,' but flood losses are largely acts of man."[8] Repeating that message for a lifetime, he became known as "the father of floodplain management."[9]

Building on White's message, hydrologists William Hoyt and Walter Langbein published *Floods* in 1955, pointing out that historic floods in 1903 ranked among the largest, yet the damage they caused in constant dollars was less than in relatively minor flooding nearly five decades later in 1949. Smaller floods were causing more damage because continued floodplain development outstripped protections being added through dams and levees.

Flood-prone development continued to exceed the federal government's flood-control efforts in a pattern growing persistently worse. Federal budget examiner John Hadd in 1965 recognized that development incentives were created by the dams and levees, writing, "Many current projects are in reality land development schemes." While dams and levees reportedly reduced damages 3 percent per year, floodplain development increased by virtually the same amount.[10] A lot of money was being spent to simply run in place, if that.

An executive order by President Johnson in 1966 marked a turning point, though perhaps only in public policy archives, with federal endorsement of floodplain management at the highest level. In the spirit of LBJ's Great Society, and with intent to launch an expansive new era of government efficiency, the directive recognized the need to reduce rates of ongoing floodplain development that otherwise made the job of reducing flood damage impossible.[11]

Through the 1960s a persistent Gilbert White pushed his case further to an academic and professional audience of planners and engineers. The public, however, remained uninterested while Congress, without compunction, entrenched itself ever deeper in concrete, lavishing funds on dams and levees serving politicians who were reliably lionized for "bringing home the bacon" in repeated triumphs of the "iron triangle" of pork-barrel politics. In this three-sided juggernaut, local developers backed their chosen politicians, the politicians appropriated funds for the Army Corps of Engineers, and the Corps delivered flood-control projects for the local developers.[12] The critical lubricant of money changed hands with each rotation of the cycle, provided the three parties remained wedded without compromise to the dominant, if flawed, vision: by spending enough money and applying enough engineering, we can control nature, or at least appear to be doing so to the taxpayers, local officials, media, and, well, everyone. Almost.

Fortunately, analysts looking deeper produced a sequence of reports through the decades following LBJ's directive and documented the need to better address flood vulnerabilities. Yet every effort to limit development faced a multipronged challenge: short-term economic incentives drew builders to flood zones, and then publicly funded dams and levees made those zones appear to be safer than they really were, leading to yet more development in harm's way, followed by greater damage and bolstered justification to build even more dams and levees. It was an exquisitely functioning, perpetually rewarding, impregnable political machine benefitting all who were on the receiving end, and it continued to grind out dams, levees, flood relief, and new floodplain development for decades.

Early initiatives to enact zoning as a floodplain management strategy—aimed at curtailing the rush of flood-prone development with its inevitable damages-to-come—were confronted by a cabal of well-heeled, well-armed opponents including not just the expected local power elite of real estate agents, bankers, and developers in communities all across America, but also vociferous property-rights advocates who went far beyond respect for the bedrock American belief in land ownership. This antimanagement cabal resisted any limits on what an individual or developer wanted to do, regardless of public cost or hazards heaped onto neighbors. Even among advocates of land-use controls, the practical challenges of mapping floodplains and crafting acceptable regulations remained formidable.

Ambiguities of the government's approach to flooding extended to the Tennessee Valley Authority, which was best known for its extensive system of dams built for flood control and hydropower, instituted while America battled the Great Depression and continuing after it in a spin-off storm of economic imperatives. However, staff at the TVA included not only the dam builders who made the agency at once renowned, adored, and despised, but also lower-key, little-known natural resource professionals striving for reforms that still rippled out from the early-1900s Progressive Era of public policy that went the whole way back to Theodore, rather than Franklin, Roosevelt. Those planners and engineers regarded nature and human efficiency broadly within a framework aiming to benefit the generations to come and not tied to political constructs such as the iron triangle. Gilbert White dryly recounted years later, "A Tennessee Valley Authority, after sobering experience with investment in flood control works without substantial reduction in total flood hazard in the area, began to look into alternative means." Among those looking was an American visionary, unsung with his calculated lack of flamboyance. TVA engineer James Goddard saw that lasting flood protection was impossible without addressing land-use issues, and he doggedly pushed TVA to aid local communities in limiting floodplain development.[13]

However, the obstacles barring his path toward responsibility for decisions about where we build and live were typically insurmountable. Swimming against powerful currents, Goddard had the presence of mind to test various semantics surrounding this minefield of public policy and concluded that the least objectionable yet accurate name for a regulatory approach to land use aiming to reduce flood damage was *floodplain management*. This became the enduring phrase for the sensible goal of recognizing where new development should best be located to avoid hazards and curb public costs. Under this practical paradigm, the floodplains, which were discrete and limited parcels of land under jurisdiction of local governments, rather than the floods, which were inevitable climatic events of global scope, would be "managed."[14]

With the principles of White and Goddard seeding new ground and expanding at the turbulent political interface of policy and popularity, the Rivers and Harbors and Flood Control Act of 1966 increased the Corps' scope of work to include "floodplain management" and modestly aimed to improve mapping and information for local communities so they could limit new building on floodplains if they chose to do so, though this remained a big and seldom surfacing *if*.

Supportive reports followed, including the National Water Commission's *Water Policies for the Future* in 1973. Then *Principles and Standards for Water and Related Land Resources Planning*, by the temporarily influential federal Water Resources Council and adopted by President Nixon, reached further, establishing guidelines intended to advance nonstructural alternatives to dam building, including land-use rules. Entirely on board as no other chief executive ever was, before or since, president, engineer, and rivers enthusiast Jimmy Carter followed up with executive orders for planners of federal projects to avoid development of floodplains and to promote natural solutions.[15] Chief among these federal initiatives, a flood insurance program incorporated goals of keeping undeveloped floodplains as open space (see chapter 8).

However, in spite of federal and academic endorsements, floodplain protection and zoning failed to gain much traction relative to the government's established programs that were, literally and figuratively, more concrete in nature. Between 1970 and 1988 only 3 of 170 Army Corps projects reflected nonstructural flood-control methods such as wetland acquisition or floodplain protections in lieu of dams and levees.[16] Difficulties included a dearth of federal funds for the unglamorous and controversial nonstructural work. This chronic austerity contrasted starkly with multimillions still dedicated to dams and levees with zero or nominal local cost-share, all putting earth movers and building contractors to work in congressional districts where money, freed by political support, was cyclically and explicitly pumped directly back into political campaigns, agency empires, and local businessmen's pockets.[17]

As the 1990s escalated toward their politically and culturally stormy conclusion, the revolution of thought supporting floodplain management resonated with a multitude of academic, scientific, policy, and agency professionals including some within the Army Corps who knew, full well, the limits of the old build-it approach. But resistance to land-use controls or management persisted from construction and real estate interests and from a constituency that objected to any challenge of individual rights or property-owner prerogatives.

In this context, and with sweeping significance to political process in a looming age of heated partisan antagonism, combining vested-interest money with conservative worldviews became a strategic playbook for opposition not just to flood reforms, but also to the entire gamut of environmental initiatives and to a host of other progressive movements crossing the American political canvas in the decades that followed.

A NEW KIND OF PLANNING

While Gilbert White extolled floodplain management to a professional audience in water resources, Ian McHarg's persuasive voice as chair of the Department of Landscape Architecture and Regional Planning at the University of Pennsylvania brought floodplains into a wider planning context.

Voice may be a bit of understatement here. With a captivating if not incendiary brogue from his native Scotland, and with fearsome impatience dating to his fearless paratrooper days jumping out of airplanes over Nazi territory at the height of World War II, McHarg embodied a lifetime of frustration at what he called the "land rapacity and human disillusion" of suburban sprawl and other environmental transgressions.[18] Where Quaker Gilbert White remained unconditionally reserved, humble, and academic, the charismatic McHarg emerged from his early life experience as a minority but bombastic challenger to the status quo. Then, bigger than life, the professor caught the wave of environmental consciousness that built toward Earth Day 1970, and he rode that wave with the temperament of an angry bear. Aiming his rage at environmental abuse and injustice wherever he saw it, and laying out the rationale for good planning, backed by convincing narrative and captivating graphics presaging an entire cartographic industry that eventually trailed in his wake, McHarg published *Design With Nature* in 1969. This lavishly illustrated volume took land-use planning beyond the work of government officials and beyond the influence of development industries and excited an audience of aspiring young planners and designers seeking solutions to the newly defined "environmental crisis." I happen to know this because I was one of them.

McHarg's approach sensibly and thoroughly plotted natural features on maps and overlays whose visual patterns enlightened planning—a graphic leap that may seem mundane in the dazzlingly chromatic GIS world of the twenty-first century. But in its time, McHarg's flashy, colorful, hand-drawn overlays of map upon map revealed fundamental processes of nature—including floods, natural drainage systems, soils, geology, and more—in a novel way and elevated them above traditional planning imperatives that had earlier focused on the facilitation of development.

Revealing both the works and limits of nature itself, McHarg showed that planning could become no less than a new and practical form of art in the highest sense, vitally relevant to the fate of the environment and to people's communities.

Though McHarg's gospel and its application by eager young planners would encounter limits imposed by an entire culture of growth aligned against them, the promise of good planning was nonetheless buoyed by hopes and aspirations in the 1970s that society could collectively change its ways, and giving rivers room to flow was one of the touchstones of McHarg's "design with nature." Thus, when this new crop of planners analyzed both limitations and potentials of land, floodplains were among the first places to be set aside for protection, followed by wetlands and steep slopes—all rich in natural endowment but challenging or prohibitive for building and infrastructure.

But despite growing attention to the merits of floodplain management, flood damages continued to grow. In the muddy wake, misfortune was explicit, graphically photogenic, and heartbreaking. During such suffering, critiques regarding precaution and self-restraint in land-use decisions were flatly unwelcome, if not considered callously insensitive to people's misfortunes. Or, if not blatantly resented, any critique of how we have become so exposed to avoidable hazards was forgotten among descriptive and gripping narratives of loss. As seasoned journalists, Tom Knudson and Nancy Vogel broke through the usual ceiling of sympathy after floods and wrote poignantly in their pathbreaking exposé on flooding in 1997, "mercy" rather than "scrutiny" follows a flood.[19]

That durable response has been, by sensible measures of compassion, justified up through the Kentucky flash floods of 2022, when thousands who lived on floodplains and who had not enrolled for flood insurance were rendered homeless after torrents of runoff devastated the narrow valleys of Appalachia, long-handicapped economically, hydrologically, and politically by boom-and-bust cycles of strip-mining for coal and the flow of its dollars through local power structures. Visiting the impoverished and newly recrippled region, a benevolent President Biden clearly, sensibly, and hopefully sustained the traditional "mercy" theme and proclaimed—though without evidence—that after such disasters we somehow always "came out stronger."[20]

However, it had become increasingly clear that incentives for floodplain development through dams, levees, and disaster-relief payouts ran directly counter to a rising profile of governmental goals to truly reduce flood damages and costs. Speaking to a conservative readership, even the editors of the Raleigh, North Carolina, *News & Observer* recognized the folly of subsidies that served both developers and victims in flood zones. In a piece descriptively titled "Awash in Tax Dollars" the editors wrote, "The allocation of hundreds of millions in taxpayer dollars has led the federal government to undermine what state officials have been trying to do for decades—discourage development in coastal areas that are vulnerable not just to hurricanes but to heavy storms of any kind."[21]

Building the floodplain management case to an audience of resource professionals, biologist John McShane of the Environmental Protection Agency combined Gilbert White's perspective about minimizing flood damage with growing revelations that floods are essential to rivers' health, writing, "Floodplain management is not just 'flood control' but encompasses two co-equal goals—reducing the loss of life and property caused by floods and protecting and restoring the natural resources and functions of floodplains."[22]

Floodplain management theory and policy had evolved impressively, though without much on-the-ground progress to show for it. And worse, in what has become a classic case of the law of unintended consequences, federal disaster assistance provided reverse incentives to individuals and communities by paying for the destruction without requiring mitigation or, as floodplain management specialist David Fowler of the Association of State Floodplain Managers described it, "The federal government pays for damages in municipalities that don't regulate their floodplains while communities with sound flood management do not need the disaster funds and get little federal assistance. Bad behavior by local governments that allow development on floodplains continues to be rewarded."[23]

Gaining clarity with their broadening message, persistent advocates of floodplain management realized that the key to addressing flood problems was to alter the behavior of people rather than that of rivers. So simple yet so revolutionary, this view recognized that we humans have a free will and, when given adequate opportunity, can make changes, while the rivers will follow the laws of nature no matter what we do.

Protecting floodplains as open space was—and remains—especially amenable to a land-use-zoning approach because, unlike other "natural" hazards such as wildfire in the West, or earthquakes rupturing across vast seismic landscapes, or tornadoes touching down randomly throughout broad belts across the Midwest and East, floods along rivers directly affect only discrete and predictable waterfront acreage—just 7 percent or so of America. To go somewhere else on the 93 percent that's not floodplain might sound like a reasonable guideline for anyone building a home. Clearly not all that land is good for housing, but a lot of it is better than the lowlands where flooding has the potential to destroy houses and even drown people.

The management alternative requires keeping undeveloped flood-prone land as open space so that floods wreak less damage. However, the path for actually doing that was riddled with obstacles ranging from the cultural to the physical, and as a result, floodplain development continued almost as if the builders, backers, and victims were oblivious to flood hazards altogether.

GAINING MOMENTUM

Catching the current of the times, the Corps stepped beyond entrenched roles with cement and rebar in 1974 by publishing a landmark document, *Flood Plain—Handle with Care*, recognizing the importance of "nonstructural" land-use approaches. Then, in one of the preeminent cases of using natural landscapes instead of structural bulwarks to combat flooding, the Charles River project near Boston graduated from concept to reality. Charismatic Rita Barron of the Charles River Watershed Association resisted plans to build an obtrusive, wildlife-killing levee system along her local river and in 1977 urged the Army Corps to instead purchase 8,400 acres of floodplain wetlands for natural storage of 50,000 acre-feet of water, equivalent to a medium-sized reservoir. This cost $10 million instead of the levees' $100 million. The Corps agreed with Barron, bought the wetlands, and floods in 1979 and 1982 proved the strategy to be sound.[24]

But then the push for alternatives withered, and flood-related responsibilities other than big structures shifted to the Federal Emergency Management Agency. Corps historians Jeremy and Dorothy Moore described the fizzling reforms at the Corps by explaining the lingering appeal of dams and levees: "Structures had been used for generations and their costs and benefits were well understood. Their physical presence instilled a source of security. Their effects were permanent and, with periodic monitoring, predictable throughout the life of the project. . . . By contrast, nonstructural measures kept people away from the water, rather than water away from the people. They employed unfamiliar and nontraditional activities like zoning and flood preparedness, which require personal involvement, and they called for individual sacrifice, such as paying for flood insurance."[25] *Flood Plain—Handle with Care* all but disappeared even from archival collections.

Filling the gap that had opened when the Corps backed off from its barely born land-use management approach, several states adopted their own programs to limit new development on floodplains. Wisconsin—at that time still reflecting progressive leadership going back to Governor and then Senator Gaylord Nelson as well as fiscal hawk William Proxmire—led this pack, hiring a capable young water resources engineer, Larry Larson, to administer a new state law restricting flood-zone development. He and other floodplain managers faced the problem of federal programs being less restrictive than those in some (but by no means all) of the states, thereby undermining precautionary strategies otherwise gaining momentum. So the state managers urged federal officials toward better partnership in flood-risk reduction. As informal chair of this fledging group of state employees, Minnesotan Patricia Bloomgren penned a notice in the program for the American Planning Association's annual meeting in 1977. Staff from not just an expected few, but a surprising nineteen states eagerly answered her call. This informal caucus kick-started the Association of State Floodplain Managers.[26] Fast-forward: by 2023 the organization had grown to thirty-eight chapters and 20,000 professionals as the nation's leading advocate for floodplain management.

THE GALLOWAY REPORT

Boosting the cause of floodplain management after the Mississippi's "Great Flood" of 1993, a blue-ribbon White House–appointed Interagency

Floodplain Management Review Committee of thirty-one experts released *Sharing the Challenge: Floodplain Management into the 21st Century*, or more succinctly, the "Galloway report."[27]

Brigadier General Gerald Galloway had risen through the ranks of the Corps for thirty-three years after graduating from West Point, also earning a master's degree in engineering from Princeton, a public administration degree from Penn State, and a PhD in geography from the University of North Carolina, followed by a prestigious resume including dean of the Industrial College of the Armed Forces and dean of the academic board at West Point, where he served as the first head of the Department of Geography and Environmental Engineering. There he navigated, among other paths, the cutting edge of global warming with its effects on national security—ominous but little known in that era. With equally impressive service on the ground, Galloway had been commander of the Corps' Vicksburg District in Mississippi and a member of the presidentially appointed Mississippi River Commission, where—more than anywhere else in America—the Corps' proverbial rubber hits the road in terms of flooding. For years the general "did what Army Engineers do," as he described that chapter of his past to me in a 2021 interview.

Long before that, Galloway's father, an Army Corps engineer himself, had passed books by Gilbert White down to his son, prepping him to later become friends with the renowned professor. "To say that I became a disciple of Gilbert White might be an exaggeration, but not by much, and I was drawn especially to the idea of alternatives to how flood problems had been approached, back to early days of the Corps. One thing you learn in military planning is to look for various ways to solve a problem. You never know when you might have to go to Plan B. Or C. Or D. For the best approach, it's important to talk to everybody, and I put that principle to work in the context of water resources."

Thus, heavily armed with an open mind, the general listened to scientists and to everyone he could engage, including advocates at the Environmental Defense Fund who were typically on the opposite side of debates with the Corps. When the need for a strategic path through the catastrophic outcomes of flooding confronted the Clinton administration squarely in the face during the 1993 "Great Flood" of the Mississippi, Katie McGinty, chair of the White House's Council on Environmental

Quality, was encouraged to telephone Galloway and to offer him the job of piloting that journey.

If anything, the Army is an organization of top-down protocol, and respectful of that, Galloway was concerned about support from the brass. So he stepped forward with caution. McGinty responded with her own insight, reminding the general that a direct line from the president of the United States, as commander in chief of the military, indeed met a key requirement in "chain of command." Further, she assured the general that the president wanted the best independent report possible, without bias toward the past or toward political influence. Galloway recalled, "I knew that the Corps had changed, and that it was increasingly receptive to non-structural solutions dating to President Carter's Water Resources Council ten years earlier, and before. So here, in the wake of the greatest flood, some of us saw an opportunity to align with approaches that many of us knew were essential."

Galloway's report recommended the expected levee reinforcement where infrastructure had failed to protect Mississippi basin communities, but leaving the Corps' deeper history of dams and levees respectfully behind, the document called for levee setbacks as corrections to earlier placements that had denied rivers their needed room-to-roam. The report recommended not more impoundments—as in the Corps' past ambitions since 1936 to plug virtually every feasible damsite on every sizable river in America—but rather improvement of dam operations, for which critical analysis showed many opportunities to improve catchment and distribution of floodwaters.

All of that may have been expected in any responsible report, but bucking the tide in greater ways, the Galloway report emphasized effective land-use rules for new development plus incentives to relocate homes and businesses away from floodplains, and especially to vacate structures that had been flooded repeatedly. The general's prospectus called for restoration of natural floodplain functions and even the abandonment of agriculture where "feasible." Recognizing that dangers had been grossly underestimated, Galloway endorsed expanding the "regulatory floodplain" in order to address hazards beyond the perceived 100-year zone. Presaging a greenway movement to come and painting a future vision that had scarcely been thinkable to most who were embedded in flooding problems, the treatise advised that "many sections of communities where flooding has been a way of life" be converted to "river-focused parks and recreation areas as former occupants relocate to safer areas on higher grounds." *Sharing the Challenge* called for a fundamental change in attitude, stating, "Floods are natural repetitive phenomena."

Floodplain manager David Fowler remembers the report as a "huge step for floodplain management and one whose ripples became waves that are still moving the Army Corps toward nature-based flood-risk reduction." Working as the floodplain manager for Milwaukee at the time, Fowler recalled that the report personally caused him "to begin thinking outside the box culvert" and that "in time, the city's department turned from an engineer-centric organization to one that addressed flooding, stormwater, green infrastructure, and climate resiliency in an integrated way."[28]

The Galloway report cautioned against what is curiously known in the insurance industry as the "moral hazard" of flood response whereby the individual's incentives to avoid damage are undercut by government payback for losses, all leading to less local initiative for sidestepping hazards. The document supported floodplain restoration, not as a virtuous afterthought, but co-equal with protecting property and saving lives. Ranking the natural value of rivers on the same plane as safety and economy marked a pivotal gain for the cause of floodplain management and for the health of whole ecosystems on which the Earth depends.

Following this report, and through the 1990s, "mitigation" for flood damage, rather than flood "control," continued to receive unprecedented political backing and emphasis. The Clinton administration reinforced initiatives for acquisition and relocation in what disaster analysts called a "new era in floodplain management and mitigation policy."[29]

Riding the tide of reform, recommendations of the Galloway effort were vigorously enacted and pursued. Federal programs protected floodplain acreage at a scale never before realized. Record numbers of flood-damaged homes were bought and moved or retired, leaving cottonwoods and sycamores to germinate and grow where doomed investments had recently stood so vulnerably.

But then, beneath the general's feet, the political ground shifted yet again, and Galloway found himself between one era that was not fully past and another that had not fully dawned. Or, as the general explained in stark political terms during our interview, "Following Newt Gingrich's conservative invasion of Congress in 1995, and the resistance that accompanied it, everything changed, and the push for new alternatives mostly went away."

In the decade following 1993 the Federal Emergency Management Agency continued allocating funds for relocation and open-space restoration, but at a low level relative to other flood expenditures, and initiatives of the Galloway report were increasingly stalled. In 2021 the retired general reflected, "Local officials wanted results right away—things they could see. And in their minds, what they could see still meant concrete." To establish the report's "river-focused parks and recreation areas as former occupants relocate to safer areas on higher grounds" was not a vision much shared in the politically influential real estate and home-building industries, even though one can argue that implementing Galloway's recommendation would catalyze a boon to new development, which, for both quality-of-life and economic reasons, thrives on the dry-side of new riverfront parklands and their greening of formerly devastated and otherwise permanently blighted neighborhoods. But in that turn-of-century era saturated with political strife, when right-wing ideologues marched in lockstep and when bipartisan support became poisoned by the polarization that political fundamentalism breeds, a reasoned approach to redirect federal flood priorities eddied out like mud-stained flotsam in a receding pool at the edge of the Mississippi. Development in flood zones again accelerated.

Following severe Midwest flooding in 2011, the Corps tried to resurrect the Galloway report's guidance, but those efforts "did not receive traction," according to Scott Spellmon, deputy commanding general for civic operations of the Corps at the time and later promoted to chief of the Corps in 2020.[30]

Though in a class by itself, the prestigious Galloway report joined the ranks of other farsighted government documents that called for better floodplain management but, in the end, were largely ignored when push came to political shove. Yet, embedded therein were succinct ideas whose time—by any sensible measure—had come, and whose relevance remained forever timeless.

"We kept going back to it," the general recalled, adding that "both FEMA and the Corps instituted long-term changes. But they just weren't enough, and still aren't." With a pause that seemed to indicate the opportunities forgone, Galloway added, "On top of all that, we now have climate change. A lot of elected officials still deny it, but just look out the window." Indeed, our interview closed as a siege of hurricanes roared to its autumn climax across southeastern states, bringing floods to ill-prepared communities, many of them with political leadership in denial of what was explicitly occurring.

Nationwide, persisting weakness in floodplain land-use regulation was noted by a Senate task force pointing out the paradox of the federal government paying for flood damages but being politically shackled whenever advocating for meaningful limits to hazard-prone construction.[31] Through it all, the key solution was resoundingly simple: enact better floodplain zoning.

THE ZONING CONNECTION

Let's back up to some of the basics of zoning—a practice deeply embedded in the heritage of land-use and floodplain concerns but nonetheless controversial and still being attacked on the conservative margins of public policy.

Zoning harkens back to ordinances passed by local governments as early as 1916. Nothing new in the 1970s, 80s, or 90s—and certainly not in the 2020s—zoning to explicitly protect floodplains from development has been enacted by some communities for decades. In 1945 the Brandywine Valley Association in Pennsylvania pioneered as America's first watershed protection organization, and its efforts led to one of the first ordinances to regulate floodplain development. Taking this idea for a larger spin, the Tennessee Valley Authority picked up on visionary James Goddard's work with the agency and in 1953 launched a cautiously titled Community Flood Damage Mitigation Assistance Program, ultimately persuading thirty-eight local governments to adopt floodplain zoning—no small feat in the politically scarlet southern Appalachians or, for that matter, anywhere. Yet zoning efforts mostly failed to expand during dam building's boom years of the 1950s and 60s with their lingering hubris that engineers and governments could reliably control floods.[32]

In spite of compelling reasons advanced by planners and public safety advocates, contrarian forces for both profit and private rights typically oppose land-use regulations. As hydrologist Luna Leopold concluded of detractors in the early days of floodplain management, "What is wanted at local levels is a program that will permit unhindered development of the flood plain through federally constructed works that reduce the risk of flooding. . . . Local interests oppose zoning."[33] That view hadn't changed much over the decades when floodplain management specialist Dennis Mileti in 1999 addressed the lack of effective zoning and reported that without "strong mandates" backed by the federal government and states, "most local governments will continue with business as usual."[34]

As Leopold said, any discussion of land-use regulation confronts the American axiom of "having the right to do whatever I want with my land." However, zoning supporters argue that such an absolute right does not exist in either an ethical or legal sense because, without reasonable regulation, one landowner's actions can usurp the rights of others. In this fluid realm, ultimately requiring an adjudicated sense of balance, courts have upheld the authority of local governments to impose limitations on floodplain development when applied in nondiscriminatory ways and when the actions of one owner could diminish or impinge on the rights and property values of neighbors; also when private decisions result in public hazards or undue costs borne by all. Well into the twenty-first century, friction over zoning persists to one degree or another, yet floodplain regulations have become widely accepted across most of the United States.

Decades of case law confirm that virtually all building can be banned where flood flows are most dangerous or harmful. A structure or artificial fill in one place might aggravate hazards by displacing floodwaters and pushing them onto neighboring properties. Beyond the most hazardous areas, zoning can bar building on land subject to even minor flooding, and rules can require elevation of first-floor levels above expected flood crests. Codes can require special building materials such as marine-grade plywood, plus flood-wary practices such as bolting buildings and mobile homes to foundations to prevent them from floating away and wrecking other property.

The rationale for zoning reaches beyond housing and typical land development. Pollution from materials stored or dumped on floodplains, for example, can cause health hazards. Industries along floodplains of

Virginia's James River have spilled toxic waste directly into floodwaters affecting communities of half a million people.[35] In Appalachia, 100 coal ash dumps and sludge ponds of powerplants have been parked in high-risk flood zones. Public damage of $1.2 billion occurred at just one of those sites when flooded in 2019.[36] Needing an effective regulatory program that extends beyond zoning, 2,500 toxic chemical sites lie on floodplains nationwide, each a time bomb with flooding as its fuse.[37] Within six cities, 6,000 former industrial sites were found with high risks of flooding where toxins were likely present, potentially causing chemical exposure in hundreds of thousands of residents, many of them economically disadvantaged.[38]

Along this line of vulnerability, agribusiness corporations have located industrial hog farms with wastewater lagoons that poison the air and water for miles along floodplains of North Carolina and other states where zoning allows them, which is almost everywhere. So consider what happens: in 2016 Hurricane Matthew flooded ninety-one hog facilities plus thirty-six poultry factories holding millions of trapped chickens and turkeys. During Hurricane Florence in 2018, pig feces, urine, blood, E. coli, and salmonella spilled from 100 flooded animal wastewater lagoons to foul southeastern rivers, communities, fisheries, estuaries, and public beaches. Instead of the ounce-of-prevention by zoning, a multimillion-dollar taxpayer buy-out was required after the damage was done, all without any guarantee that a new generation of hog operations won't reappear on flood-prone land.[39] Local or state governments fail to adequately regulate industrial hog and chicken facilities, yet the North Carolina Department of Agriculture in 2022 had no qualms about asking FEMA for tens of millions of federal dollars for postflood cleanup of dead animals and disease-bearing waste that could have been avoided by local and state initiatives, had those governments taken responsibility with zoning and siting rules.[40]

Still another rationale for floodplain zoning—and one that's growing with the climate crisis—aging public infrastructure and increasing severity of storms will require emergency releases from upstream dams, putting homes in the path of those flows at risk. Zoning can limit housing where dam managers will need to hastily dump floodwater to avert dam failures. Such zoning for disaster prevention rarely exists, and new homes continue to be built where floodwaters will need to be released.

While a select group of communities had enacted floodplain zoning before 1973 without federal incentives, the vast majority had not. After that, upgrades of the National Flood Insurance program led to most vulnerable municipalities adopting floodplain zoning to one extent or another (see chapter 8). The good news is that thousands of communities have now enacted at least some regulations to control floodplain development. Most of the 22,000 municipalities enrolled in the insurance program have zoning that bars at least some development in the most hazardous areas. The bad news is that typical zoning still allows widespread building across much of the 100-year floodplain, and it greenlights development on virtually all the floodplain above that level, which includes a lot of land destined for flooding as the climate clouds thicken.

Through a history of legal challenges and contentious debates, the right of local governments to regulate flood-zone properties has been maintained, even in spite of a 1992 case, *Lucas v. South Carolina Coastal Council*, when the Supreme Court, which had turned conservative with appointments in the 1980s, ruled 5–4 along political lines that a coastal zone regulation constituted a "taking" and thus required compensation for the person who wanted to develop but had been barred from doing so. In other words, the highest court ruled that to limit one landowner from endangering another landowner or public resources, the one causing the problem has to be paid. Such obligations could extend to paying off a landowner who simply states an intent to develop. Law professor Rutherford Platt described the *Lucas* decision's ensuing double standard: the federal government is increasingly called upon to pay for disaster damage owing to shortsighted land-use decisions, but "government at all levels is increasingly impotent" to curtail "unwise developments in hazardous locations."[41] Dennis Mileti of the Natural Hazards Center likewise described the post-1992 legal atmosphere as a "chill on land-use regulation" when rules are enacted "for disaster mitigation purposes."[42]

Hazard zoning may have been cooled by the *Lucas* decision, but by no means has floodplain zoning been stopped. To comply with the court's ruling, zoning maps can identify areas where at least some development is allowed along with places where it's banned. Water policy expert Jon Kusler surveyed the legal landscape and found that communities enforcing sound floodplain restrictions sustain a solid record of court victories whenever those municipalities have been sued for over-regulation. Conversely, local governments allowing developments that result in damage to neighbors have mostly lost in court.[43] Yet, expensive legal battles can lurk in the way of communities that try to meaningfully limit floodplain use.

Backers of zoning faced interest groups such as the California Realtors Association, whose spokesperson announced at a state senate hearing on disaster prevention that his group would "find restrictions on development repugnant."[44] Nicholas Pinter of the University of California pointed out that in many people's minds "private property rights trump even modest limitations on floodplain development."[45] With varying degrees of interest, ability, and outlook, local government officials weigh these and other arguments for and against zoning wherever flooding occurs.

THE PATH FORWARD

The floodplain management approach has grown, from a discussion of planners and scientists in Gilbert White's career, beginning at the height of the Great Depression, to widespread zoning of floodplains, to public greenbelt corridors of open space, and onward to flood-control systems of the 2020s based on "natural processes." Yet this proactive, precautionary, and management approach to flooding remains underused while a warming climate promises greater storms and higher waters threatening everybody at an accelerating scale that far outpaces our regulatory grip on the problem.

It seemed that we had turned a corner back in 1966 with President Johnson's directive for floodplain management, or in 1972 when Hurricane Agnes unveiled the weakness of past approaches to control floods, or in 1993 with the Mississippi's Great Flood and the wisdom of the Galloway report. But one has to wonder how much has been accomplished. Like the fight against global warming, launched back in 1988, the effort to reduce flood damage has been thwarted, obfuscated, and spun into distracting narratives benefitting some while victimizing many. As Rob Moore of the Natural Resources Defense Council warned, "For every house that we allow to be privately built in flood areas there will be a family that we'll publicly have to help in the future."

Watching this issue for several decades, Andrew Fahlund of the Water Foundation recognized, "Whenever we are faced with catastrophic floods, there is a brief moment when people pay attention and realize that more must be done about the management of land use on floodplains, but then the crisis passes and little is accomplished. Trying to reform the way we manage flooding has been avoided even by most conservation and water management groups because it requires specialized expertise, engagement at a river-basin scale, and various political skills rarely found under one roof. Yet, the climate crisis is going to require people to tackle issues like land-use regulation of floodplains, and a lot of challenges need to be overcome."[46]

The complexities of those issues emerge in California's response to growing threats of high water. After extensive debate, the state legislature in 2007 passed the Central Valley Flood Protection Act, widely regarded as a robust step toward better regulation in flood zones. The law requires that further development in urban or suburban flood-zone communities of 10,000 people or more cannot proceed without adequate levees or other protection to a 200-year-flood level rather than FEMA's ineffectual 100-year standard. But will a 200-year limit be adequate given future floods of the global-warming era? "No," state floodplain manager Mike Mierzwa frankly admitted, adding that communities have the option of requiring greater protection.[47]

Moreover, limits on development of California floodplains that are rural—a vast majority of the flooding acreage—are not required by the state law. Regulation there continues under local land-use rules, in most cases influenced by the National Flood Insurance program and its focus on solutions such as elevating structures above the 100-year flood. The state has done little to fill the void left by ineffective local regulation and misdirected federal incentives. Thus, even in flood-prone, flood-aware California, the focus of mitigation has remained on improving levees around urban areas and doubling down on structural defenses rather than on preventing rural development of floodplains or relocating threatened homes and farming operations away from danger.

While zoning across the US has surely precluded a lot of building that would otherwise have occurred on the most-hazardous "floodway" portions of floodplains, the big picture shows a battle, if not a war, whose outcome is unknown, if not being lost. According to NOAA in 2014, more than 40,000 acres of natural and agricultural land in designated floodplains were converted to development between 2001 and 2006 in coastal counties alone.[48] That's a grim statistic for anyone caring about an issue that has been debated ever since Gilbert White was in his twenties.

And brace yourself for this one: home construction near the coasts has increased fastest *within* the most flood-prone areas.[49] Let me say that again: in spite of all efforts to restrict building on floodplains, the rate of development *inside* coastal flood zones has been greater than *outside* them. Census Bureau data revealed that the number of people living on 100-year coastal floodplains climbed a remarkable 14 percent between 2016 and 2000 while the population in areas *not* likely to flood increased only 13 percent.[50] Of course, some areas are far worse than the average. Sea level is rising. Floods are getting worse. With our foot tromped on the accelerator, the vehicle of flood-prone development is headed straight toward a wall. As Dennis Mileti reported in *Disasters by Design*, "The United States had been—and still is—creating for itself increasingly catastrophic future disasters."[51]

A reflective eighty-six-year-old emeritus professor in 1997, Gilbert White recognized success in floodplain management but also key disappointments, and he soberly concluded, "This is very humbling to me. I saw this problem, and a lot of people saw this problem, a long time ago. Why didn't we have a greater influence?"

To address his question, it's important to recognize complexity, so let me put on the optimist's hat, and with acknowledgment that floodplain management confronts an entire culture of free-wheeling development that's disinclined to yield an inch, let's note that efforts to keep floodplains as open space have, indeed, been impressive, with progress clearly evident since the first steps were taken in the 1950s. Congress leveraged goals forward with flood insurance requirements in 1973 and with floodplain regulations that have been accepted, to one degree or another, in most areas, and remarkably so in some places. Consider Nashville—a blue city in a very red state—which has enacted effective limits on floodplain development. Zoning allows new building in floodways under "almost no conditions," according to Metro Government floodplain manager Roger Lindsey. Builders on floodplains beyond the high-hazard area must elevate first-floor elevations of new homes, not just 1 foot, but 4 feet above 100-year-flood levels. I asked, "Do you still get pushback from property-rights groups claiming anybody should be able to build however they want?" Lindsey responded, "No. I think those days are behind us."

Likewise, Knoxville requires strict standards on building up to the 500-year-flood level—far beyond minimum requirements of the National Flood Insurance program. These southern cities in deep red states have shown that floodplain zoning can certainly be effective, given the local will to make it work.

Though flood-destined properties continue to soar in value—especially along the coasts—recognition of greater risks finally seems to be growing even in the real estate industry. Laura Craft of the investment firm Heitman noted that a 2019 survey of real estate property managers indicated most had not considered the problems of increasing flood vulnerabilities. But surveys only one year later indicated a shift, with managers becoming concerned about "market-level risk" from floods.[52]

Going back to Professor White, many people all along the way in this story have known that their times demanded transitions if not transformations. With the effects of a heating climate and greater floods, an increasingly compelling case can now be made to protect floodplains and to direct new development elsewhere. That seems like the least we should be doing. The faucet for the overflowing bathtub can be turned off. But even if every acre of open space on today's floodplains is safeguarded, millions of houses have already been built in flood zones. Many are subject to repeated flooding followed by relief payments, rebuilding, and flooding again. For these, the National Flood Insurance program has taken center stage.

8   The Insurance Connection

COMBATING RISK WITH INSURANCE

The boom in dam and levee building through the 1900s failed to adequately prevent losses from floods, and consequently disaster payments ballooned beyond control while widespread suffering haunted the future. Farsighted policy analysts and concerned members of Congress eventually recognized what virtually everyone knows when confronted with the similarly catastrophic prospect of their home burning down: people at risk need to buy insurance.

Insurance appeals as a free-market solution to flooding, popular in politically conservative circles and also beyond them. But unlike house fires, floods on floodplains arise inevitably over time, wreak damage in riverfront corridors extending many miles, cripple economies of whole regions, and rank four times more likely to damage houses than do fires.[1] Moreover, floods have been getting worse, damages rising, people's ability to recover falling. Well informed, the insurance industry knows all this, which makes flood insurance unattractive in their business model and therefore an unaffordable backstop for homeowners, especially those with low incomes. So in an effort to curb taxpayers' escalating outlays for disaster relief, and to sympathetically address victims of flood losses, Congress adopted a program of federally sponsored and subsidized flood insurance.

Just for clarity, here, disaster relief and flood insurance both offer payments to flood victims, but the approaches distinctly differ. Taxpayers collectively and solely fund disaster relief. Government aid is made available to essentially any flood victim when a disaster is presidentially declared, with an average of $7,400 per property flooded, and with a cap of $33,000 per flood event, as of 2022.[2] This aid is not intended to cover big-ticket items in rebuilding or relocation, but principally to keep victims safe and sanitary in disasters' wake. Additional help can come in low-interest loans from the Small Business Administration, even for applicants that are not businesses, and from other government programs. Insurance, on the other hand, requires yearly premiums from the recipient, and then the program pays out more per flood incident than does disaster relief. The median net payment for insurance claims in 2017–22 was $33,870, and the maximum was $250,000 per flood, notably with no penalty or limits on repeat damage and payouts.

The insurance effort is ostensibly self-funded, but as we'll see, it's not. Enormous subsidies have kept a well-intentioned and presumably economic program on life-support. Beyond that, federal flood insurance has become a classic illustration of the law of unintended consequences. Let's consider the history and the outcome of this intriguing government program that, with both vigor and compelling arguments, has been both praised and maligned.

Back when reasonable bipartisanship characterized what now seems like a nostalgic age of workable politics, World War II to 1995, the "management" approach to flood protection—as opposed to structural efforts to "control" or stop floods—had gained political momentum through economic and common sense. With management rather than strictly flood-control goals in mind, the first attempt to apply insurance as a tool came with congressional action in 1956 under the helm of popular, practical President Eisenhower. But with dams still being built at a furious pace and with support for alternatives lukewarm at best, the insurance initiative languished, an unfunded idea whose time had not yet come.[3]

Pressed by those looking deeper for solutions to the intractable problem of increasing flood damage, Congress appointed the Task Force on Federal Flood Control Policy, which in 1966 issued *A Unified National Program for Managing Flood Losses,* recommending a broad approach including land-use regulation in hazard areas. Gilbert White chaired the group and predictably wrote, "Flood damages result from acts of men. Those who occupy the flood plain should be responsible for the results of their action."

Worth underscoring, the Task Force recommended bolstering the insurance idea—not even a workable program then—but presciently warned that subsidized insurance should not become an incentive for more floodplain development. "A flood insurance program is a tool that should be used expertly or not at all. Correctly applied, it could promote wise use of floodplains. Incorrectly applied, it could exacerbate the whole problem of flood losses. . . . To the extent that insurance were used to subsidize new capital investment, it would aggravate flood damages and constitute gross public irresponsibility."[4] With those words in 1966, America should have considered itself warned.

Taking to heart the concept of using insurance not just as help for people who need it, but also in the spirit of the Task Force's call that this program could be an incentive to "promote wise use of floodplains," Congress in 1968 amended the National Flood Insurance Act, strengthening it with what has been called America's first major "nonstructural" approach to reducing flood losses—a cause for celebration by many.[5]

Like any political remedy, the legislation struck a bargain, this one in classic carrot-and-stick fashion. It offered subsidized insurance to people already living or invested on floodplains, provided, however, that local government take regulatory measures to avoid worse damages to come. This sensible goal was to be met by requiring zoning that would halt new development in flood-prone areas. Better described as a "carrot-and-carrot" approach, the subsidized insurance and directive for floodplain protections were intended to meet people's needs and also aim society toward better use of floodplains overall, together offering both private and public benefits the whole way around. First under the Department of Housing and Urban Development, the program was passed off in 1979 to the Federal Emergency Management Agency (FEMA), now in the Department of Homeland Security, better known since 2001 for its high-profile job protecting the country from terrorists. Which, by the way, has proven to be far more effective than protecting us from the flood dangers that have cost and killed far more Americans than has terrorism.

The 1968 act's authors optimistically and sensibly envisioned that vulnerability to floods would decrease over time because new flood-prone development would essentially be banned while old development morphed, by attrition, from vulnerable buildings to public greenways along rivers, or to privately owned open space, or at least to buildings that were effectively flood proofed and no longer posing economic and public safety liabilities.[6] At long last, America's nettlesome problems of flooding might be destined for solution. However, with insurance as with life, the devil hid in the details and in this grand plan's evolving roadmap, which would lead to places unknown and, as it turns out, unintended.

With a reinforced framework for flood insurance barely launched after the 1968 legislation, the 1972 Hurricane Agnes Flood raged through the Northeast, graphically confirming that traditional flood-control efforts—dams and levees—can be neutralized if not perversely turned hazardous with structural failures when it simply rains for three days. Meanwhile a tally of homeowners who had joined the insurance program fell somewhere between disappointing and pathetic, and tens of thousands of victims—nearly all, in fact—were uninsured. Failed levees in Wilkes-Barre, Pennsylvania, displaced 100,000 people but—with unwarranted faith in their levee's thin pile of rocks separating urban neighborhoods from the mighty Susquehanna River—only two homeowners there had bought insurance. Two. Redoubled efforts to gain subscribers, to reform failing elements of the program, and to expand its goal of minimizing vulnerability all followed with the Flood Disaster Protection Act of 1973 and its renewed efforts to get serious about avoiding flood hazards.

LEVERAGE AND ENTICEMENT

In order to upgrade the insurance program, the 1973 act made coverage mandatory for mortgages backed by federal deposit insurance, which includes bank loans that for most people are prerequisite to building or buying a house. Congress reasoned that the government, and not just the homeowner, incurred risk with floodplain investments because federally backed mortgages end up in the government's lap when homeowners suffer floods, fail to make payments, and default on their loans. This brilliant bit of legislative leverage had precedent with a little-known, post–World War II policy blocking Veterans Administration loans to applicants in flood-prone areas.[7] Flood insurance under the 1973 act remained optional for people owning their home and not needing a mortgage. Municipalities also remained free to not join the program, but then their residents would not get federal mortgage insurance for new homes in flood zones. Flood victims who don't buy insurance can continue to receive relief with presidentially declared disasters.

As a professional planner deep in the trenches of county government in 1973, I can say—and I think I speak for many here—that we were thrilled with anticipation that the bolstered insurance program would lead to better use of floodplains. More than that, we thought that the dual incentives and limitations promised no less than a new chapter in the history of land use in America. After years of trying, planning in the McHarg mode of "design with nature" was staged to achieve sharp limits on development of floodplains—clearly one of the most hazardous types of landscape for locating a house or business and also the worst place to develop if the timeless functions of nature are to be respected.

But exactly where new construction was prohibited and what type of building was to be allowed within flood areas both remained subject to influence from a formidable quartet of lobbyists knocking on Congress's doors: the US Chamber of Commerce, banking industry, real estate organizations, and National Association of Home Builders. All came unified in fighting restrictions limiting new construction on floodplains, and all were ready and able to spend a lot of money lobbying Congress and the administration.[8]

FEMA in 1986 revisited flood insurance efforts with its *Unified National Program*, reiterating that "avoidance of development in high hazard areas is the preferred approach." But with thirteen years for the act's impetus to cool off since the motivating crest of Hurricane Agnes, and now under the Reagan rather than Nixon administration, the agency ceded that construction in some flood zones may be appropriate in light of the "public interest" or lack of a "suitable alternative." Such rhetoric opened wide the loophole door. Seemingly gun-shy in the face of critics from the building industries—if not blatantly aligned with them—authors of a 1986 FEMA Inspector General report similarly explained agency reluctance to set high development standards by noting that "mitigation policies" aimed at reducing future damage carried the unwanted perception of "negative economic impact because they discourage development in hazard-prone areas."[9]

In 1989 a national review committee, again chaired by floodplain management veteran Gilbert White, soberly reported, "On balance, progress has been far short of what is desirable or possible, or what was envisaged at times when the current policies and activities were initiated. . . . Losses to the nation from occupancy of riverine and coastal areas subject to inundation are continuing to escalate in constant dollars."[10]

The reason was familiar to anyone who had pondered rising flood damages throughout the whole previous century: as the review committee stated, most losses "can be attributed to increased property at risk." In other words, development in flood areas continued to grow while effectiveness of flood control and regulatory programs failed to keep up or, for that matter, to get adequate traction at all.[11]

The insurance program's signature duality and genius—subsidies for people suffering damage in return for local government assurances that floodplains would be zoned to reduce or eliminate future damages—had been eroded as the intentions for land-use rules were watered down, requirements softened, and enforcement neglected. In other words, the talented lobbying cabal for development had tightened a firm grasp on political process and quietly fueled a property rights movement proclaiming the individual as supreme over the community or, for that matter, over any *other* individual. This movement gained fervor, as General Galloway reported in chapter 7, after the Republican takeover of Congress in 1995 and thrived under the party's "Contract with America" and its manifesto opposing government regulation, government engagement, and government itself. Backed by right-wing think tanks and bolstered by legal cases strategically aimed at conservative judges, the property rights movement put land-use reformers into full retreat and demanded that property owners be paid if regulators curtailed them.

Disaster analyst and land-use legal expert Rutherford Platt noted that relative to earlier goals of the flood insurance program, the "conspicuous trend" in the 1990s was "decreasing emphasis on land-use planning and management in flood hazard areas." He maintained that the federal government paid out more and more in relief and required less and less in regulation, conflicting directly with original flood insurance commitments that participating communities "adopt adequate floodplain ordinances with effective enforcement provisions consistent with federal standards to reduce or avoid future flood losses." Original requirements for zoning were replaced with a saccharine goal of "positive attitudes" toward floodplain management.[12]

Far from a ban on development of floodplains, FEMA requirements turned out to be as minor as adding a single course of cement blocks to the height of a foundation, thus raising the first floor by as little as 8 inches above the projected 100-year-flood level, however inadequate that standard was. The rules became a game that developers quickly learned to play, billing their clients for whatever adjustments were needed.

A House bipartisan natural disasters task force in 1994 dug deeper and reported that insurance payouts, especially for repeated damage, were counterproductive to the fundamental purpose of the insurance program, stating, "People are encouraged to take risks they think they will not have to pay for."[13] The task force reported that the goals of the insurance program had failed to be met and, more incriminating, that hazards were actually *increased* by shielding risky investments with subsidized insurance. That was precisely what the House of Representatives' *Unified National Program for Managing Flood Losses* had warned about in 1966. Authors of the textbook *Floodplain Management* later concluded that by limiting property-owner risk through insurance and disaster payments, the insurance program "has perpetuated the development of river environments."[14]

Apart from the disappointing regulatory side of the program, a bipartisan congressional task force on disasters in 1994 addressed the federal role in paying for flood damages, and it feared dependency on government handouts. It supported improved insurance but called for retreat from increased federal spending and for a return to "supplemental" roles for the government in order to "encourage all members of society to bear their fair share of the costs of disasters."[15] Not just a sound-bite from Republicans predictably applying the brakes on public assistance, this message was amplified by bipartisan disaster analysts recognizing in 1999 that the "gradual expansion of the federal role has been accompanied by a growing sense of entitlement to federal disaster victims."[16] Yet federal expenditures and the flood victims' dependency on them continued to grow.

In spite of the insurance program's goals, FEMA payments for flood relief (not including insurance payouts) in the wake of continuing disasters expanded during the program's formative years, then soared by 23 percent between 2000 and 2010, then escalated further with intensifying storms the following decade. Though Congress had intended to curb disaster outlays, the program didn't come close. Disaster payments topped $460 billion from 2005 through 2019 from FEMA and other federal agencies, all in 2019 dollars.[17] Deficits loomed larger, floods grew stronger, reform receded deeper, with systemic breakdown of the insurance program a real possibility.

Even with the mortgage limitations, not all municipalities enrolled for federal insurance, and many still have not. In August 2021 flash floods killed twenty people in two Tennessee counties that had refused to enter the program and enact the zoning it required. Free to speak his mind after retirement, former director of the federal insurance effort Roy Wright explained that communities shunning the program typically do so because of aversion to restrictions on building in dangerous flood zones.[18]

In practice, disaster relief with or without a community being enrolled in the insurance program is broadly applied; even snowplowing has qualified for "disaster" aid. Federal relief is granted, of course, because genuine human suffering merits help within a sympathetic society, but also because of political optics around people in distress. When governors request emergency support after a flood—and they have few reasons not to request aid—no president wants to look like an ogre by refusing it.

Thus, in spite of original intentions for the insurance program, new floodplain development continued and was spurred onward by insurance subsidies. The test of time has shown that government-sponsored incentives to rebuild on the footprint of damage outperform conflicting government enticements to relocate upslope toward safety.

Unforeseen by all—except perhaps real estate agents keeping a keen eye on this ball—building boomed along flooding coastlines. The insured value in flood zones along Gulf and Atlantic shores increased from $7.2 trillion in 2004 to $10.6 trillion in 2012, skyrocketing onward in the following years.[19] Enormous piles of wealth were invested in lowland properties of extremely high risk, but now that risk was moderated by a guarantee of subsidized federal insurance covering $250,000 per house and $100,000 for contents, per flood, available flood after flood in 22,000 eligible communities.[20] The payments represented substantial aid to victims and delivered repeated reimbursements—over and over with each additional flood—for damage that added up to extremely high sums (see chapter 9).

Administration of the insurance program pointed toward FEMA doing just enough to meet minimum legal requirements of previous congresses and courts. In other words, compromises permitting developers to continue building on floodplains showed caution, not so much about the threat of floods, but rather about a litigious system sharply tilted—as *Lucas v. South Carolina Coastal Council* confirmed—toward individual rights rather than community values. The permissiveness of the flood insurance rules, and the risky building that it promulgated, all pointed to the power of players who thought they stood to make less money if the flood insurance program actually cut flood losses, as was intended.

Zoning and building codes of local communities are needed to underpin the insurance program, and these were analyzed by FEMA in 2022, with distressing news: floodplain regulations in thirty-nine states received the lowest grade, what in school we'd call an F. Many of these, including Louisiana, North Carolina, and Pennsylvania, ranked among the most flood-prone states. Top scores for effective codes went to California, New Jersey, and New York. The deficient codes in the thirty-nine states exposed people to "a dangerous, costly and unnecessarily high level of risk."[21]

To be fair, FEMA's original and dominant mission was to distribute federal relief after disasters. Then Congress gave the agency the edgier job of overseeing regulation of developers, which always attracts a heavy dose of influence. But not long after its formation in 1979, FEMA was roundly criticized for becoming home to proportionally more political appointees versus professional civil servants than any other federal bureaucracy. Leadership positions were gifted by administration officials to friends who had done little to deserve the job other than deliver political support in the previous election cycle.²² Independent journalist and historian of hurricanes Eric Jay Dolin described the agency as a place "where good friends or supporters of the president got plum jobs even if they had no experience or skills that would recommend them to the posts."²³ This was alarmingly true when Hurricane Andrew in 1992 followed on the heels of President George W. Bush gifting his former campaign director—widely regarded as a political figure clueless about disasters—the top FEMA position.

Of course, many if not overwhelmingly most FEMA staff are dedicated public servants doing the best that circumstances and overlying politics allow, and clearly not all FEMA appointments were politically driven. President Clinton, for example, elevated the agency to a cabinet level with its full vetting and approval process, welcomed scrutiny, and appointed as director James Lee Witt, who had extensive professional experience and drew from it effectively in his tenure as chief. But short-lived, the cabinet status was lost when President Bush rolled the agency into the Department of Homeland Security, which was preoccupied with terrorism. More recently, President Biden's appointment of Deanne Criswell to lead FEMA in 2021 was based on a distinguished career in disaster management, including degrees in public administration and homeland security followed by municipal work in Colorado, for FEMA, and as New York City's commissioner of emergency management.

Even when watered down as the insurance program's regulatory requirements had become, they still brought the topic of land-use planning to the forefront of municipal matters across America and onto the agenda of many local officials typically more engaged in paving a township road or hiring a new policeman. However, permissive loopholes if not open gates allowed for continued building, provided certain and too-often trivial floodplain limitations were met.

None of these disappointments had to be. Municipalities, in fact, may enact stronger zoning than is required by the federal program and are encouraged to do so through a smart, merit-based FEMA point-system that justifies lower insurance rates.²⁴ But confronting their own issues of influence, and limited by budgets and inadequate staffing, local governments rarely seek to exceed minimum requirements.

To understate the level of disappointment here, the outcome was not what Congress or the insurance programs' backers in floodplain management and related hydrologic disciplines had intended. Instead, the results served to validate historian John Barry's succinct observation, "When you mix science and politics, you get politics."[25]

MAPS AND MANIPULATION

As a key to the entire insurance and zoning enterprise, floodplain mapping has to be done well. People need to know the boundaries of floodplains in order to determine if their land or home is likely to go under water. But mapping remains an art as well as a science, it relies on history as much as engineering, and its uncertainties tempt political intervention. At its core, the mapping effort typically looks backward rather than forward for guidance—rarely a good bet. In chapter 6 we considered the difficulties of accurately mapping and projecting the levels of floods. The insurance program and its regulations inherited every one of those problems and raised even harder issues about floodplain maps and the decisions stemming from them.

FEMA has used the 100-year floodplain as the area subject to insurance regulations, even though, as we've seen, that level has proven to be chronically inadequate, and extremely so in many cases. But to avoid regulations that might be regarded by some interests as overly strict, this 100-year zone is further subdivided into two parts: "floodway" and "flood fringe." The floodway is where depth of water and velocity of current create explicit hazards. The flood fringe is also expected to flood within the 100-year cycle, but the damage potential there is considered less severe.

Within floodways (high-hazard areas), homes and other significant developments are not usually allowed, although planner John LaVelle of Lycoming County, Pennsylvania—on the front lines of regulation—explained that building is not so much banned, as one might expect given the original motivation for the insurance effort, but made "less likely" owing to requirements. In flood fringe areas, which typically account for more acreage within the 100-year zone than does the floodway, regulations permit new homes if they're built to a "high regulatory standard" incorporating special construction techniques and materials. To collect insurance, homes incurring damages exceeding 50 percent of the house's

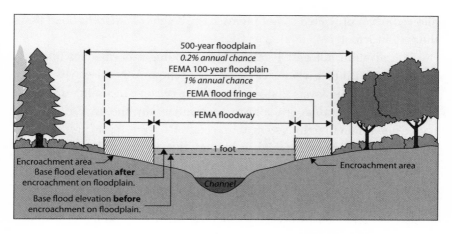

This diagram shows the extent of a floodplain likely to be inundated once per 500 years and the smaller area likely to flood each 100 years. The 100-year floodplain is further divided for flood insurance and zoning purposes into a "floodway" having the highest risk of damage, and a "flood fringe" that will also likely be flooded on 100-year cycles. The "encroachment" areas indicate where new development might be allowed but will cause flood levels to rise. Expected flood levels shown on the FEMA maps are often exceeded because of inadequate data, ongoing watershed disturbance, and the increasing floods of a warming climate. Illustration credit: "One Foot Increase in Base Flood Elevation—Section View," Federal Emergency Management Agency.

preflood value (however determined) must elevate to 1 foot above the 100-year flood. The cost of these typical "flood proofing" measures is more than that of conventional construction but not generally prohibitive; case in point being the single row of cement blocks mentioned earlier. The deeper the water during the 100-year flood, the higher the cost of mitigation. Some homes meet requirements by elevating as much as 10 feet, perched in what might kindly be termed an odd architectural style, though the elevation can also be architecturally disguised, at least in theory.

One of FEMA's most controversial policies allows subsidized insurance for new development in the flood fringe area (which typically includes the majority of the regulated acreage) if buildings are sited on fill that's piled up until it tops the 100-year-flood level. This earth-moving, water-displacing enterprise is cogently called "fill-and-build" and has been allowed when projected to not increase the height of adjacent flood flows more than 1 foot.

Fill-and-build not only legalizes but encourages permanent filling of floodplains with dirt whenever they're not temporarily filled with water, all leading to flood levels that rise yet higher because incoming water has to go elsewhere in this zero-sum game. Incidentally, the fill also destroys native floodplain functions such as percolation of runoff into groundwater tables and nourishment of wildlife habitat, thus undermining goals of letting rivers function more naturally. But the main rub with fill-and-build is its aggravation of flood hazards for neighbors.

Consider, under this controversial rule, that the fill piled up for one new home might not be projected to increase nearby flooding appreciably (though every little bit could count), but enough fill to accommodate two new homes might do it. What about ten or twenty new homes, approved one at a time? This is often the trajectory of subdivision along undeveloped floodplains. Larry Larson of the Association of State Floodplain Managers recalled that federal regulators had originally intended to address these scenarios of cumulative fill-and-build causing flood levels to worsen appreciably. But then the rule-writers "dropped the cumulative idea," and so those aggravating effects have largely escaped consideration, even when their harm to neighboring properties is likely and substantial.

The fill used for fill-and-build is literally dirt cheap, allowing for a lot of development and floodplain encroachment. Also, consider that flood crests of the modern era routinely overtop 100-year expectations by not just one, but multiple feet. In the end, many fill-and-build permits become gambits inviting flood damage, they ruin nature's floodplain services, and they put neighbors all around at greater risk. A fleeing flood victim might step out the door and into a current of water whose swiftness owes entirely to a neighbor's fill and new home that pushes floodwaters away from it and into the critical escape route.

Here's another troubling variable with fill-and-build and with the 1-foot rule for elevating houses: hydrologic findings regarding proposed developments are often calculated by consultants paid for by developers eager to downplay risks. Most local governments can't afford expert reviewers of their own, and in many cases cannot even afford a professional planner, and so the developers' numbers go unchecked by municipal officials who might be uninformed, or don't have time to check or even to look, or are pressured by the builders, or for whatever reasons are disposed to join with them in undermining regulations. In the rare case when local governments conscientiously try to zone floodplains *more* restrictively than FEMA requires for insurance, overt influence from developers is the norm. I can personally attest to this, having spent eight years as a county planner in a flood-prone county at the eye of this contentious storm.

Some jurisdictions do opt to require first-floor elevations that exceed the 100-year-flood level. Pennsylvania requires a foot and a half; Wisconsin, two feet; and Nashville, four. However, most municipalities go with the minimum.

Making accountability difficult, the FEMA maps are cumbersome, or perceived as such, for citizens to comprehend. Seeking remedies in 2020, the First Street Foundation, which tracks floodplain issues, released its Flood Factor tool to make the maps more accessible.[26] The Association of State Floodplain Managers also has an online tool to calculate flood risk.[27] Yet the complexity of it all remains cryptic to many landowners.

To meet even minimum requirements, FEMA and state officials recognize the need to update thousands of flood maps, but the agency faces backlogs because of the high expense and chronic underfunding by Congress.[28] All in all, shortcomings in mapping result in "100-year" storms occurring far more frequently than 100 years and in people buying properties without knowing the risks. Meanwhile urban flooding from intensified storms in unmapped neighborhoods has shocked cities as notable as New York, where in 2021 an unexpected deluge horrifically drowned forty-six people when water rushed into basement apartments and prevented escape. Let me again point out a blind spot here: if forty-six people were killed in a terrorist attack, the response would have been far more notable.

At least the Association of State Floodplain Managers is trying to target the problem: the group urged Congress to fund mapping of 2.3 million miles of streams where flood levels have not been plotted. Many of these mostly small-stream basins continue to be developed and flooded without land-use rules or even rudimentary flood-hazard knowledge.[29]

Having what is likely the most perceptive historical knowledge of flooding issues, the Floodplain Managers' co-founder, Larry Larson, succinctly concluded, "When the National Flood Insurance program was created in 1968, there was a belief that floodplain management regulations would reduce development in high-flood risk areas. We now know that the expected reduction has not happened. Instead of stopping floodplain development, the program has told us *how to build there.*"[30]

Larson further noted that the Army Corps of Engineers had originally recommended use of a 500-year floodplain as the regulatory standard for flood insurance rather than the smaller 100-year zone. He also recalled that both the 100-year cutoff that triggers insurance protocols and the 1-foot requirement for fill-and-build were arbitrary, and chosen in spite of inadequacies pointed out by informed critics.[31] For that matter, back in 1967 the *Proposed Flood Hazard Evaluation Guidelines for Federal Executive Agencies* defined floodplains as areas inundated not by a 100- or even 500-year flood, but by a 1,000-year event.[32] In the quick shift from hydrologic analysis to political application, a thousand years was trimmed to a hundred, and I'm reminded again of historian John Barry's observation regarding the milkshake of science and politics.

Then too, as we saw in chapter 6, much of the data underpinning the flood maps is less than a century old, and resulting inaccuracies subvert insurance efforts. As early as 1994 the Association of State Floodplain Managers reported that some 31 percent of damages being claimed were located *outside* the mapped 100-year floodplain. This meant that substantial new development was allowed in flood-prone areas without any location or construction limitations.[33]

Fast-forward twenty-three years and not much had changed. A Department of Homeland Security Inspector General report in 2017 still found that many FEMA maps did not reflect actual flood risks. Only 42 percent of the maps were up to date. Maps of 1970s vintage predated many record-high waters that followed. Though FEMA commendably spent $200 million in recent years updating maps, more work awaited; 40 million people resided in what should have been considered 100-year floodplains but were not. That's three times the number of people indicated on FEMA maps.[34]

Taking this overweight bull by the horns, the State Floodplain Managers in 2020 issued *Flood Mapping for the Nation* with comprehensive recommendations for better mapping and for including areas of "residual risk" that have not been plotted because damage potential was regarded as "low." This owed not to a low likelihood of flood probabilities and not to any low degree of objective hazard, but rather to low levels of existing development. Yet those lightly developed areas are being built up all the time, many of them quickly becoming developed without mapping to show home buyers the hazards and without flood insurance requirements such as rudimentary zoning.[35]

In 2022 FEMA reported that insurance payments made to cover flood damage totaled $74 billion, 30 percent of which lay outside of mapped 100-year floodplains—virtually the same as what the Floodplain Managers reported in 1994.[36] Further, since it takes seven years on average to produce a map, congressional requirement for reassessment every five years means the documents are arguably out of date before being released.[37] Meanwhile Congress continues to appropriate inadequate money for updating the maps. Regarding the traditional 100-year threshold, improvement even to a 500-year floodplain as a standard of risk compares poorly to the Dutch, who have coexisted with floods for a long time and—wiser for the wear—use a protection zone based on the 3,000-year highwater event.

The bottom line is that flood mapping has for years underestimated risks and also failed to recognize the space that rivers need to overflow and to be—in any real sense of the word—rivers.

FAILURE TO DISCLOSE THE INEVITABLE

A critical purpose of the mapping and the zoning that stems from it is to simply warn people, including prospective buyers, if property is flood prone. Even the inevitable can be unexpected, unconsidered, and undisclosed.

I was once tempted to buy a cabin along Oregon's Rogue River where the real estate agent assured me that "they built a dam upstream so it will never flood again." That didn't quite pass either my eyeball test of what I saw or my sniff test of the negotiation. In checking flood maps in the county planning office, I found that the cabin of my dreams in fact lay squarely in the floodplain, just awaiting an atmospheric river with my name on it. I declined to buy, and whoever did likely had no idea of the flood hazards.

Simple disclosure of flood risk is arguably the easiest, cheapest, and least controversial remedy one might imagine toward reducing flood losses. FEMA generates critical information that's valuable as a warning to people, but in most cases neither the agency nor anyone else is obligated—or, one might assume, disposed—to share those warnings with real estate shoppers. FEMA cites privacy privileges as the reason for not making flood-damage information available to prospective buyers or— let's face it—to tomorrow's flood victims and to the public at large. Privacy? Really? Taxpayers are paying for these maps. They're prepared because of a public need. But what they reveal is withheld by restrictions involving privacy? Why is a deed search to guarantee authenticity, safeguard against fraud, and provide safety to the buyer not also an invasion of the seller's privacy? Real estate agents could easily surrender flood information to their house-shopping clients, and determined shoppers can take their own initiative in gleaning essential data from FEMA's floodplain delineations, but Nashville Metro Government's Roger Lindsey tactfully noted, "Teaching home buyers or the real estate community to use the floodplain maps has proven to be a slow process."

Since 1996, and probably earlier, agency staff have recommended disclosure of flood probabilities.[38] Federal disclosure requirements have likewise been considered by congressional members, but gauntlets of opposition have thwarted every effort to inform the public about what it needs—and has a right—to know. In October 2021 FEMA announced that it was "considering" recommending a disclosure law; however, that hopeful communiqué didn't mention congressional roadblocks, where real estate industry lobbying likely loomed as a frigid Everest in the political milieu of the day.

Apart from disclosure that remains hidden behind the FEMA flood-risk-curtain, there are no state-level requirements for disclosure of flood hazards in real estate transactions and rentals at all in 21 states.[39] In 27 of the 29 remaining states, critical information is surrendered to buyers only after they make an offer on a house. Considering how much deliberation goes into making a firm bid on, say, a $400,000 house (US median in 2002), this "Oh, by the way" accommodation to ethics via disclosure typically comes too late. Even then, all that the buyers might see is a checkmark in a box, easily overlooked, especially when checks in boxes typically indicate passing rather than failing grades. Undoubtedly speaking for many, home shopper Gloria Horning of Pensacola lamented, "You don't find out that the house is going to flood until you sign on the dotted line."[40] For renters, only eight states address relevant risks. Georgia requires notification only if property has been flooded three times in the past five years—a curse for anyone, let alone the renters' cohort who can least afford repeated losses. Often on the side of consumer protection but manifestly flunking in these regards, California doesn't require disclosure if flood risks are less than 1 percent in most areas, leaving out everything beyond the 100-year-flood zone, which includes a lot of floodable property. The area subject to disclosure was decreased further to a likelihood of half of one percent in the state's Central Valley, which happens to be where most of the flood-prone land is sold.[41]

Motivated by hurricane experience affecting many people, Texas and Louisiana passed effective disclosure laws, but to see what typically happens when states try to require disclosure, consider Virginia, where bills failed twice. The rejections occurred after the state real estate association argued oddly in favor of ignorance regarding flood hazards rather than having the buyer be exposed to what the association derisively called "information from a nonprofessional." Puzzling at many levels, the real estate group's critique, as reported on National Public Radio, didn't seem to recognize that the real estate agents, themselves, were the ones who would be providing the information.[42]

One can only assume that those who oppose disclosure requirements fear that the property won't sell—or will sell for less—if people are informed about what they're buying. This time I'll paraphrase historian Barry: when you mix ethics with political influence, you get political influence.

To sidestep the congressional impasse, the Association of State Floodplain Managers recommended a model disclosure measure for enactment by states.[43] A recent bright spot, New Jersey in 2023 passed by unanimous vote of its legislature a disclosure law with agreement and cooperation of the real estate industry—a credit to efforts on both sides.[44] But at the national level, Rob Moore of the Natural Resources Defense

Council summed up the sorry status of disclosure: "If we deny people information, we shouldn't be surprised that they make mistakes."

FALLING SHORT

For perspective on what was intended with the flood insurance program versus what we got, consider that in 1997 FEMA stated a goal of reducing annual flood costs by $1 billion. In fact, costs skyrocketed.[45]

In coastal flood zones across the South and East, building rates soared in spite of the insurance program, in spite of threats of higher premiums, in spite of forecasts for increased flooding, in spite of the likelihood of investments being stranded by a heated climate and fractured economy shattering in the wake. A 2019 survey by the research group Climate Central found that coastal areas at highest risk, with a 10 percent chance of flooding in any year (a 10-year floodplain), were seeing their highest rates of home construction in a decade. Local officials typically cheered the boom, anticipating next year's property-tax revenues but not the bankrupting tab for next year's floods, which takes us back to the familiar theme of denial. Also to the theme of letting the federal government—otherwise anathema in real estate matters—absorb a large share of the risks and costs of the most risky and costly development that's happening.

Struggling to simply blow a whistle across the inevitably flooding coastal floodplains, no less than thirteen federal agencies attuned to global warming in 2018 jointly warned, "Many individuals and communities will suffer financial impacts as chronic high tide flooding leads to higher costs and lower property values."[46] But few people—or at least public decision-makers—seemed to care in the lubricated party atmosphere that accompanied the real estate boom.

Here's another failure of the insurance program that, one might say, is also troubling from a social-justice standpoint: the federal subsidies make no distinction between homes of primary residents and structures used as second homes or retreats, including lucrative vacation rentals. Lavish trophy houses built solely for leisure and with extreme damage potential consume large amounts of insurance program payouts, all while essential and modest full-time residences barely get paybacks needed to keep people housed. Commendably stepping in the proper direction, FEMA in 2021 proposed new rate schedules whereby the most expensive houses will be charged more for coverage.

The insurance program's multiple challenges result in staggering subsidies paid every year to merely keep the enterprise afloat. FEMA flood insurance paid out $89 million per year in 1980–84, but $1.6 billion annually in 2016–21. In an age of bankrupting disasters, the program owed $21 billion to the national treasury, not counting $16 billion forgiven by Congress previously in a stratagem to evade limits established in earlier years by a Congress explicitly committed to not running up the debt. The flood of red ink keeps flowing with no end in sight.[47]

Somewhere in these discussions the question may have come up, why do people continue to rebuild in areas known to flood? Part of the answer to that is now clear: they build there because federal taxpayers cover a significant part of the damage when floods occur.

REFORM OR RETIRE

Efforts to reform the flood insurance program have long drawn from positions historically supported on both sides of the political aisle, though *historically* is a key qualifier. Republican lawmakers in the past have typically been critical of domestic spending and ostensibly favored free-market solutions such as people paying for insurance. Democrats traditionally back public safety, socially conscious payment formulas, and environmental protection. But ideological support for the insurance program fell victim to self-interested business lobbying that outspends both of the left and right political traditions and both of the platforms that have typified the two parties. Real estate associations wanted free reign without regulations, and builders wanted to build wherever people or the government would pay them to do it. A cadre of honest politicians and bureaucrats tried their best to counter those persuasions, and they deserve credit for their efforts, but a blizzard of influence rendered meaningful political philosophies down to what could only be called a farce.

Free-marketers in Congress ended up voting for subsidies, and social justice advocates ended up voting for trophy homes that devour government funds meant to help people most in need. The *Houston Chronicle* reported extensively on the issue of "influence" after the Hurricane Harvey Flood, concluding, "Attempts to fix flood insurance have been derailed repeatedly by special interests, political expediency and powerful lobbies that have poured hundreds of millions of dollars into congressional campaigns." The investigators noted that the National Association of Home Builders spent $39 million lobbying Congress, 2005–17, and that 140 real estate companies and trade groups had poured $350 million into congressional campaigns while lobbying for deliverance from sensible floodplain rules.[48] In 2022 the US Chamber of Commerce and National Association of Realtors—both engaged in long-term efforts to influence the flood insurance program—were ranked, by wide margins, as the top two groups nationwide in money spent lobbying Congress.[49] All the smart and socially conscious Gilbert Whites and Larry Larsons of the world were thoroughly outgunned, not by logic, economics, or ethics, but by cash.

Back in 1976, R. Dell Greer, HUD's director of flood insurance for the Northeast, wrote with hope that the Flood Insurance Act would establish "procedures through which communities will eventually reduce the mounting loss of life and property resulting annually from floods."[50] His inspiring vision, championed by many, seemed to have been forgotten in recent decades, leaving the *Houston Chronicle* reporters to itemize the failures: the insurance program was supposed to discourage development in flood-prone areas, but new development spreads across floodplains. It was intended to insure properties vulnerable to flooding, but 80 percent of flood-prone properties remain uninsured. It was intended to reduce disaster relief, but those costs have exploded and still increase. It was supposed to be self-supporting, but premiums don't come close to covering expenses. The journalists grimly concluded that "subsidies have helped lure generations of homeowners into properties that trap them in a cycle of building, flooding and rebuilding."

In 2017 renewed reform efforts sought to limit repeated claims and to facilitate buy-outs. But Politico reported that the National Association of Realtors and the American Bankers Association warned that a "regional foreclosure crisis," reminiscent of the 2008 housing bust and its Great Recession, would somehow reoccur and somehow undercut all of us if taxpayer support of homes with excessive flood insurance claims were phased out. The Association of Home Builders' CEO deftly pivoted when criticized about subsidies to the rich, respinning the accusation with populist antiregulation rhetoric: "Any policymaker that would try to tell his or her population that they can't live near water is really in for a tough row to hoe."[51] Cutting against his industry's grain, Colorado home builder Ron Jones—formerly on the Home Builders' board—sensibly asked the Politico reporter, "How many times do you need to rebuild in certain areas before you say it might be better to locate somewhere else?" Then he answered his own question from years of experience: "Most home builders will be happy to build those houses as many times as somebody will pay for them."

Still another questionable aspect of the insurance program involves FEMA's inability to enforce existing rules that require communities to take steps to decrease future flood damages, including the program's core mandates for zoning and for elevating first-floor elevations. Naomi Kalman, a University of California geographer, found that a quarter million insurance subscribers violated program rules between 2010 and 2020, accounting for $1 billion in flood claims. Virtually no penalties were issued.[52]

Rob Moore of the Natural Resources Defense Council pointed out that, even when pursued, measures to enforce compliance with land-use regulations of the flood insurance program remain ineffective. "When a community fails to enforce the required ordinances, action is rarely taken. But when it is, FEMA first puts a community on 'probation,' meaning that no new policies are allowed, and existing policy holders are fined $50. But that prevents more people from joining the program, and it unjustly punishes those who are already participating. Next, FEMA can kick a community out of the program, but what good does that do? There are no effective consequences for violating the rules, yet enforcement is essential. The whole approach needs rethinking."[53]

Reasons cited by FEMA for the infractions included claims that communities didn't know they were violating program rules, that inadequate mapping was used, and more to the point, that local governments and FEMA alike simply don't want to penalize participants. When requested to clarify its enforcement policies, a FEMA representative replied, "Most deficiencies and violations are due to a lack of awareness and full understanding of NFIP criteria and lack of technical skills in the community." The agency emphasized priorities in "resolving problems through technical assistance rather than through enforcement action."[54]

Given widespread frustrations, patience runs thin in discussions about balancing the insurance program's budget, and also about tolerating repetitive losses at taxpayers' expense. Talk about inequities: the repeats account for 1 percent of the number of insured properties but 30 percent of payout dollars for damage (see chapter 9). Constructive critics, foremost being the Association of State Floodplain Managers, call for fundamental changes beyond FEMA adjusting the rate schedule. Exposing a worsening ledger, the First Street Foundation in 2021 projected a program deficit of $20.8 billion for just one year.[55]

For the insurance program to be effective in discouraging or establishing adequate regulation of development on floodplains, gaping loopholes need to be closed. Chief among these is the allowance of fill-and-build. Another loophole would be addressed by expanding to the 500-year-flood zone the area where insurance requirements restrict development. Additional improvements would be restriction on payments to repeatedly damaged properties plus flood disclosure during real estate sales. The State Floodplain Managers have long advocated for these and other reforms. However, with frustration at the failure to enact incremental upgrades, other suggestions to altogether ax the federal flood insurance effort have surfaced.

No surprise, reservations about the program's existence date to when an aging Gilbert White, in the 1990s, was "haunted" by the question, has the program become more of a problem than a solution to flood-loss reduction and floodplain protection?[56] The sharpest critics would relegate national flood insurance to a historical footnote: an excellent idea that failed owing to special interest compromises.

Interestingly, advocates for throwing in the towel on flood insurance cross traditional political lines. Sounding a lot like classic Republicans of the pre-1995 era, Judith Kildow of the Monterey Institute of International Studies—affiliated with Middlebury College—and Jason Scorse of the Center for the Blue Economy—a "sustainability" organization—wrote, "Homeowners and business owners should be responsible for purchasing their own flood insurance on the private market, if they can find it. If they can't, then the market is telling them that where they live is too dangerous."[57] The popular *New York Times* column featuring conservative Bret Stephens debating liberal Gail Collins found both supporting dissolution of the program.[58] Even left-leaning analysts pose the question, for all its good intentions, has the federal government—subject as it is to influence—only made things worse? A free-market approach where property owners sink or swim might be best.

But others say that the old cliché really fits: don't throw the baby out with the bathwater. Suffering from floods is real to many victims, and the federal government will be saddled with disaster payments of one form or another whether or not people buy insurance. Having been centrally involved for decades and supporting corrections rather than dissolution, Larry Larson concluded that the program remains our only means to get floodplain mapping, which is the first step toward avoiding losses, and that even with shortcomings, the effort includes requirements and incentives for mitigation of damage. For all its battlefield tatters, the original mandate to reduce future flood losses endures.[59]

Bolstering those arguments, Rob Moore of the Natural Resources Defense Council patiently recognized, "The program is about more than insurance. It establishes at least minimal land-use standards for floodplains in 22,000 municipalities. It's a source of flood-hazard information for anyone willing to look."[60]

With the question of the insurance program's existence or extinction hanging in the balance, the stentorian voice of FDR may come to mind: "Better the occasional faults of a government that lives in a spirit of charity than the consistent omissions of a government frozen in the ice of its own indifference."[61]

Rising in defense of its work, FEMA in 2020 released a study encompassing 18 million buildings constructed between 2000 and 2016 and reported that municipalities with flood-safe codes avoided $27 billion in losses compared to governments lacking modern codes. Avoided property losses estimated for the 2000–40 decades totaled $132 billion. The agency's message: we're better off with the program than without it.

TRYING TO GET IT RIGHT

To address overruns in the insurance program, the Biggert-Waters Flood Insurance Reform Act of 2012 sensibly required FEMA to charge rates closer to the government's actual costs and to prioritize relocation of structures rather than rebuilding in the same flooding places.[62] *The Washington Post* editorialized in support of the rate increases: "Subsidized protection against risky behavior—like building in a flood plain—encourages people to take risks, with the inevitable losses that entails. . . . It takes some chutzpah for National Flood Insurance Program beneficiaries to act entitled to subsidies from the vast majority of taxpayers who do not live on the beach—or could never afford it in the first place."[63]

With opposition to this view principally from owners of expensive coastal homes who saw their insurance rates going up and property values headed down, Congress didn't act and, in 2014, put a hold on the rate hikes it had clearly mandated two years before. Another attempt to raise rates, surgically directed at repeatedly flooded properties and new construction, also fell to a defeat credited to the building industries.[64]

Expensive coastal houses tended to be the pivotal point in these debates, and with this book's focus on flooding at rivers, not seashores, it's worth noting that the largest deficits in the insurance program come from hurricanes' surging surfs along Gulf and Atlantic shores where sea level is rising. It's only going to get worse. In fact, hopeless. So the question might arise, should coastal damage be separated and reformed in a stand-alone program? Spotlighting the enormity of the seacoast imbroglio, Hurricane Katrina in 2005 and subsequently others pounded the insurance effort into depths of debt that began to look like a death spiral for the entire program, backyard creeks in the mountains to ocean edge at all three of our coasts.[65] Noticing in the 2000s the sharp shift from river to coastal costs, the Office of Management and Budget warned, "The NFIP is not designed to handle catastrophic losses like those caused by [Hurricanes] Harvey, Irma, and Maria."[66] Genesis of the insurance program, in fact, evolved without an inkling of rising sea level, nor the staggering extent of jeopardized wealth that would accumulate at all the trendy seashores. River and coastal flood damage are two entirely different species of disasters that grew more and more different over time; however, no proposal has surfaced to separate the two forms of flooding.

To keep the National Flood Insurance program's river of cash flowing to disaster victims who in good faith paid for coverage, Congress in 2018 passed the NFIP's seventh stopgap extension within just a few years while political debate pitted congressional members who were fiscally conservative, and wanted to slash the program, against greater numbers of representatives from both parties with ears to voters who relied on the subsidies for their investments, however misguided.[67]

With no easy call politically, the continuing flow of disaster dollars raises difficult issues of equity, incentive, and effectiveness. Help of some kind for flood victims surely remains a public responsibility in any social contract recognizing needs of the less-fortunate among us. But analysts have long pointed out that the "public subsidy of private folly" can undermine the goal of reducing flood risk.[68] A vexing dilemma persists: how can a sympathetic government help people in need yet avoid undermining the sense of caution that every individual and community should adopt in staying out of harm's way?

CRISIS WITHIN CRISES

In spite of the insurance program's commitment to cut flood losses, damages between 2010 and 2018 grew to four times those in the 1980s.[69] Pardon me, but to state this yet again, the principal reason was the increase in development in the danger zones. Chad Berginnis of the Association of State Floodplain Managers acknowledged, "The great unrealized purpose of the program was to effectively steer development away from hazard areas, and that has not occurred."[70]

Beyond the avoidable and politically plagued failure of a once-farsighted insurance program, the flood of federal dollars paid out when governors requested disaster declarations grew to resemble the opprobrious if not corrupt "pork barrel" of the bygone dam-building era. James Wright, former head of TVA's Office of Floodplain Management, wrote, "The flood control construction program of the 1930–1950 era now seems to have been replaced by an equally massive federal relief and recovery assistance program for flood disasters in the present age."[71]

Lofty goals of the insurance program appear to have been derailed, leaving a fundamental question unanswered: how can government's response to disasters be better aimed at not accommodating but rather avoiding disasters in the future?

With brighter prospects for reform in the Biden administration, FEMA in April 2021 announced another attempt at the modest and common-sense goal of revising rate schedules to address the era's timely concerns about inequities across economic and racial lines. Homeland Security Secretary Alejandro Mayorkas announced, "We are putting equity at the forefront of our work and making reforms to help our nation confront the pressing challenges caused by climate change."[72] Under FEMA's new plan, "Risk rating 2.0," people with lower-value homes would pay less while those with higher-value homes would pay more; 23 percent of policy holders would see decreases in premiums, 66 percent would see minor increases up to $10 per month, 7 percent would pay another $10–$20 per month, and 4 percent would see significant increases principally reflecting top market values of high-risk coastal vacation retreats.

Pointing further to the need for comprehensive reforms beyond the social-justice canvas of the 2020s, the Association of State Floodplain Managers in 2021 petitioned FEMA to upgrade flood-zone building standards and revisit flood-worsening practices such as fill-and-build. The rules had not been reconsidered since developers enjoyed a clear political advantage during the Reagan administration. Joining in, Rob Moore of the Natural Resources Defense Council pointed out that "FEMA has documented over and over the need for stronger land-use ordinances and has done everything in its power to promote them except for updating its own requirements, which is the only thing the federal agency can actually control." FEMA agreed to a system-wide review, opening a door to reforms such as expanding the regulatory standard from the present 100-year flood to a 500-year flood, raising the height of the requirement for rebuilding to more than the present 1 foot above expected flood level, and eliminating fill-and-build loopholes. For all of flood insurance's shortcomings, it shouldn't be that hard to fix the program. It just requires political will.

Meanwhile, for all the disappointments, the federal program in 2022 insured 5.1 million properties including 3.5 million single-family homes and 1 million condominiums in 22,000 municipalities. Though big numbers, these still left the vast majority of floodplain properties without insurance and reliant, rather, on disaster-relief payments. During the floods of 2022, for example, few victims in Kentucky had enrolled in the program. Though extremely flood prone, that state fell behind only arid New Mexico in its percentage of residents buying flood insurance.[73] With few insurance policies, and with state political leadership famously hostile to a federal presence from strip-mine controls on down, Kentucky nonetheless received $159-million-and-counting in federal relief funds as of March 2022 for just this one flood.[74]

For millions of people already living on floodplains, the National Flood Insurance program was intended to provide support replacing the relief payments of the past. But even if the insurance effort were fully successful, to truly become secure from the ravages of floods means not simply being insured and reimbursed for damages after they occur. Security comes only from moving off the floodplain and relocating to higher ground.

9 Moving to Higher Ground

LIVING IN THE PATH OF DANGER

Protecting undeveloped floodplains as open space is essential in order to avoid greater damage to both homes and natural rivers, as we've seen in chapter 7. But what about floodplains that are already developed? Other than buying insurance—which, in the best of worlds, can pay for losses but not prevent them—what can be done about homes that sit in the danger zone?

A lot of people live on floodplains, and opportunities to flood proof existing buildings, and stay there, are often elusive, expensive, and ineffective, especially for the most precarious homes with repeated damage, and especially given that future floods will be worse, and worse to an unknown degree. Our flood-control infrastructure of dams will only be effective some of the time, and the security of levees will be increasingly uncertain except for urban areas where massive investments will be required to gain reliable protection or anything approaching it.

Moving off the floodplain is the ultimate answer to the hazards of flooding, and often the only answer, and it's also the only way to effectively restore the hydrologic workings of rivers and the natural values that they can provide to all.

The Government Accountability Office in 2021 assessed the nation's climate-related relocation needs and urged Congress "to provide assistance to climate migration projects." The agency pointed out that relocation from areas such as floodplains "will become an unavoidable option" and that the status quo—spending a half-trillion dollars in disaster relief for floods, hurricanes, fires, and droughts between 2005 and 2020—has not been a good investment.[1]

Possibilities for relocation include millions of people. The First Street Foundation calculated in 2020 that 14.6 million properties pose "significant risk of flooding," and the number was projected to grow by a shocking 61 percent in thirty years if policies don't change. These numbers consider small streams, sea-level encroachment, and modern rainfall records as factors, though FEMA has not included many of those risks in its projections.

Ranging from residents to bankers to federal officials, people have underestimated flood danger to existing homes, and millions of property owners have no clue about risks they face by simply staying put. Most climate-disaster victims elect to stay and rebuild after damage occurs.[2] But the more they are flooded, the more they chose to move, and the future exodus from coastlines and riverfronts will eventually reach a level unimagined today.[3]

Worldwide, needs to relocate, particularly in response to sea-level rise, are a quantum level more alarming. Here in the US we've got Miami, New Orleans, and Galveston, but densely populated, low-elevation, coastal areas in countries such as Bangladesh include homes to hundreds of millions. In 2021 the United Nations reported that climate disasters already displace 20 million people annually—2.4 times New York City, *per year*.[4] The International Organization for Migration projected that a mind-boggling 200 million climate refugees will be on the move by 2050. That equals the entire US population in 1967. To accommodate them one would have to *invent* a United States of the Johnson administration. Though the human problems inherent in these numbers are simply not on most Americans' radar, the social, economic, environmental, and political ripples, or more accurately, tidal waves, will create a different world, far harsher than what we know today.

Any fine-tuning of these numbers—or any winnowing of them to include only the people who face the greatest hazards—will not cast doubt on the fact that many people live in the path of high water and that by any sensible calculus they need to move. Inevitable prospects of even low-end projections for both sea-level rise and river flooding call into question important aspects of American life, including the boilerplate, thirty-year mortgage and the structural framework of finance governing real estate and whole regional economies.[5]

For example, Stanford University researchers calculated that because flood risks have not been considered, houses on floodplains were "overvalued" in the 2021 market by $11,526 per house. With this being just the first snowball of an economic avalanche to descend, the study questioned the entire "stability of real estate markets as climate risks become more salient and severe."[6] Increasing the decibel level of warnings in February 2023, a group of government and NGO authors published an article in *Nature Climate Change* saying that floodplain homes could lose 10 percent of their market value in coming years—a loss affecting low-income owners the most.[7] Many homes will undoubtedly lose more. Adding complexity to the economic matrix, property being "overvalued" doesn't mean that people won't pay an inflated or ill-advised price that's ultimately regretted as a bad deal. Along that line, a study in the *Journal of Environmental Economics and Management* reported that home prices in Florida rose 5 percent during various three-year periods following recent hurricanes—not after the hurricanes were forgotten, but *directly following them*. The unsettling explanation is that wealthy home buyers are not dissuaded by flooding, and as a result, they sometimes displace lower-income people who cannot afford subsidized insurance or to recover from flood damage and end up desperately searching for shelter or becoming homeless or close to it.[8]

FEMA offers help for people who want to move, but no one has drawn up a far-reaching, comprehensive, or effective plan for relocating at the necessary scale or anything remotely approaching it. To the contrary, renowned Duke University professor of coastline geography Orrin Pilkey said, "I don't see the slightest evidence that anyone is seriously thinking about what to do with the future climate refugee stream."[9]

What to do about development that's already on floodplains, but essentially doomed, is a complicated and awkward question, though it shouldn't be. Anyone who stands at the edge of a river and sees the water rising might know to get out of the way by moving higher. But contrarian beliefs in such matters have often trumped fact through two centuries while our country enacted the charade that we could stop or control all floods through engineering and fortification.

It doesn't do much good for society if a flood refugee sells their house to the next unfortunate victim, so the great escape requires that the government buy out the owners of floodplain homes and dedicate the property as permanent open space. Hope for this option is evident in dozens of towns and with tens of thousands of home relocations already undertaken along rivers nationwide. But before considering those accomplishments, and before being inspired by them, let's acknowledge the reluctance that some people have in seeking the higher ground that promises their own security.

RESISTANCE TO CHANGE

Arriving as the water receded in November 2021, I was confronted with flood damage at house after house in Hamilton, Washington, located a few hundred yards from the banks of the Skagit River but undeniably in part of the channel when the water rises. Floods occur there at three-to-four-year intervals, with really big ones, as this was, every few decades. The aftermath could only be called grim.

Piles of flood-ruined furniture, appliances, toys, papers, clothes, and lifetime mementos lay in saturated brown-stained piles covering whole front lawns as high as windowsills of soggy houses behind them. Struggling with the futile goal of drying out, people had strung tarps over leaking roofs. The brightest colors in town—the only bright colors—were the shiny blue jobsite johnnies occupying front yards of homes whose sewage had backed up with floodwater and stubbornly resisted drainage. Other yards accommodated dated motor homes serving as primary shelters, probably intended as temporary but rapidly becoming permanent, many of them already tinted green with mold. Some homes had been abandoned, broken windows open to persistent rain, bedraggled curtains blowing in the breeze. Other houses appeared to be empty, but cars, or more likely pickups, parked out front indicated otherwise. Some rusty old trucks hadn't been driven for quite a while, or maybe just since before the

In Washington the community of Hamilton floods at frequent intervals, with large floods, such as the one in 2021, rising once every few decades and causing repeated damages and relief payments.

flood, it was hard to tell, as receding water has a way of making everything look damaged. In this and other respects it was difficult to separate ravages of time from those of recent days, but mold that had accumulated on siding, roofs, and doorframes over the years was now on green, petri dish steroids of lingering dampness. As muddy and tattered as it might have been at Appomattox, a Confederate flag sagged over one door, two thousand miles from the Mason-Dixon Line.

Ever since Hurricane Agnes in 1972 I've seen a lot of flooded towns, but Hamilton was rough. Rain was falling again, and another siege of flooding was forecast for the coming night. Life was grueling with no sign of letup.

Strolling the streets, camera slung around my neck but under my raincoat, hood up, I stepped aside now and then for a grunting dump truck, mud splattered to its windshield and the spacious bed loaded with tons of rocks sized from basketball to hay bale. Otherwise I encountered nobody out and about. Knowing that many of the houses sat empty and unguarded, and that looting would be a legitimate fear, I hesitated to knock on doors. Yet ever curious, I wanted to inquire about the flood.

At a waterlogged lawn in front of a waterlogged home, I finally met up with four men, all lacking waterproof gear, all seeming to accept the continuing drizzle as a normal part of community conversation, or maybe not even worth acknowledging. I approached cautiously, not wanting to be the annoying stranger, despised voyeur, or—for that matter—the curious reporter, however legitimate a gaggle of questions might be in my mind. But it didn't require more than a friendly greeting, on my part, for one of the men to step forward, hiding a cigarette with smoke curling up behind his back and noticing the camera bulging under my parka. Bearded, protected only partially from the elements in a red windbreaker, camo pants, camo cap, and black knee-high rubber boots, he seemed eager to talk after learning of my mission. For the sake of this narrative, I'll call him "Bill."

"The water was that deep," he offered, pointing to the porch in front of us. We chatted for a few minutes about the neighbors' various close calls and deep immersions, and he seemed to get the drift of my curiosity. "You wouldn't believe how close my house was to being flooded," he said. "It's right over there."

Grateful to have a local guide, and for his combined amiability and knowledge, I walked alongside Bill as we strolled down the street, slippery with mud. He told me it was Hamilton's third-highest flood that he knew about. A still-saturated slough of the Skagit had made an island out of the little burg, 300 souls in all, and then the runoff got serious as it rose in height and sliced with unexpected, accelerating swiftness through the heart of town, nine of ten homes on the floodplain.

We crossed a low, hardly perceptible levee that separated, by only a few vertical feet, most of Hamilton from the Skagit's less-serious floods, though the levee obviously had trivial effects on this one. I could see a hundred yards down the levee's straightaway to where two of the trucks I'd seen earlier came into view, now idling, each waiting its turn to dump a rocky cargo alongside a yellow, robot-like excavator-with-a-grapple, whose metal jaws swung back and forth as rapidly as the man at the helm could manipulate the heavy loads.

We continued, now on the lower, river side of the unimpressive local defense, walking along Bill's driveway and foot-deep ruts. The compact but sturdy house had wisely been built on top of an elevated concrete slab, and Bill pointed to a telltale mud line of the recent crest, just below the sill of his front door. The back side of his house faced an unblemished flooded forest—at that moment a northwestern Okefenokee with only the alligators and snakes missing. "That's the mighty Skagit, right there," Bill said, pointing to the inundated trees, "though it's quite a ways out to where the river usually runs. This was big," he admitted. "We have floods here."

Seeing that his house was unequivocally an island only a few days before, and an extremely small and pregnable one at that, I asked, "Do you have a boat?" I never face flood risks without one and personally had my canoe strapped to the roof of my van, just down the street, where I knew I might be using it, even later that day, to closer eyeball the floodwater backup that seemed destined to grow again.

"Yeah, I do, but it's over at my shop now." I didn't ask but wondered if Bill's shop was as precarious as his house, or perhaps more so. Struck by the gravity of the scene, which pretty much defined risk, I was impressed by Bill's casual comfort.

"I've lived here thirty years. I don't want to go anywhere. Look at it! No stoplights or traffic. The Skagit's right over there."

I like rivers. I understood what Bill was talking about. I got it.

But then, in a tone I measured as 60 percent restraint and 40 percent contempt, he added, "Maybe you've heard about a plan to relocate us."

I had, though not in quite those terms. A Seattle-based land conservation group, Forterra, had committed $1 million, free of charge, to help the town relocate willing flood victims upslope. Shelling out the cash, the group actually bought 48 dry acres to accommodate those who volunteered to upgrade. This impressed me as a generous incentive to the community and an act of kindness rare in today's world.

But Bill was clear: "It's not a relocation. It's an *annexation* to the town for people who want to build up higher. Most don't want to move. This is the way it's always been. You should talk to the mayor."

Though uncertain if I should inquire, but hurdling this boundary, as journalists must, I asked if Bill got help from government programs for flood victims. "Just family and friends," he said with evident pride. "We stick together here."

I later learned that other residents in Hamilton received $4 million in flood insurance claims between 1995 and 2020 and another $1.4 million for buy-outs and grants to the city.[10] Those numbers apparently did not reflect basic flood relief to uninsured victims—which most in Hamilton probably were—nor did they include this latest and largest flood, destined to dwarf earlier payouts by a lot. On a per capita basis, Hamilton residents have received more money than hard-hit victims of Katrina. Beyond all that, additional federal money comes from the Small Business Administration and the Army Corps of Engineers, which was helping in the current levee upgrade, and from other agencies, such as the State Department of Transportation. It had its own splitting headaches at low spots in the highway, in and out, and with overtime invoices, just to keep the road to the rest of the world open.

Bypassing all that, and thinking ahead to a new era of climate or perhaps only to the rising waters of the approaching night, I asked, "Are you ready for the floods to get worse?"

"I'm not so sure about them getting worse. NOAA says that in the 1800s there was a flood that hit 60 feet. Last week it was only 32. My house is 33. Heck, if it gets to 60, it may as well be 100." As I juggled these numbers, it was evident that we didn't need to consider a warming climate to know that the floods to come could be at least as high as those of the past.

Taking Bill's advice, and searching for the mayor, I found three men sporting orange vests and hardhats, all looking professionally municipal in front of a motor home that had become the mayor's refuge since her house burned down in an unrelated misfortune the summer before. "She had to go to town," one said, referring to nearby Burlington. "Give her a call. I'm sure she'll talk to you." I asked what was up with all the dump trucks pounding Hamilton's streets.

"During the flood we discovered a low spot in the dike, plus other damage from slumping. We have to fix it before this next flood arrives. Due tonight. So we got the Corps to declare an emergency, and with their approval and help, we rushed to hire a contractor. That's what you see going on here. We're racing the clock. I gotta go."

The mayor, I later learned, had moved to Hamilton just three years before, took issue with the previous mayor over the Forterra-related plan to aid in relocating repeatedly flooded homes, and after a contentious campaign, won election by four votes, according to news in the local paper, *Crosscut*.[11] I eventually tracked the mayor down at the fire hall. In overalls and looking understandably worn from the circumstances, she graciously offered me a muffin from a box on a conference table. Then, with evident willingness to share her view, she described getting a tour, at the height of the flood, in a friend's waterskiing boat skimming over streets where I had just walked. "A ski boat?"

"Yes."

It all seemed so hopeless, even surreal. I asked what the people of Hamilton now wanted to do.

"We're a small town but we don't give up," she answered.

"The town floods over and over again," I pressed. "What's the solution?"

"The residents are the solution," she offered. "They will not let the town die. The flood has brought us closer together. I'd go as far as to call it a blessing in disguise. We're seeing better things happening now, with the dike repair. If it weren't for the emergency, we would not have gotten such support for fixing it. People realize this town is worth fighting for. Maybe you've heard of the Forterra money that's being used to make other building sites available. That's not to replace the town. It's just an addition to the town."

"Are you concerned that floods will get worse?"

"Yes. Because of dam removals. To benefit salmon, nearby Native Americans want to get rid of the Seattle City Light dams upstream. But once we hit flood levels, those dams are managed by the Corps of Engineers for flood control. Without the dams we'll have greater flooding."

Following news of that type, I'd not heard of a tenable plan to eliminate the 530-foot-tall Ross Dam, widely considered essential for powering everybody's lights and life support throughout Seattle. Letting that pass, I clarified, "What I'm talking about are higher floods that result from global warming. How will you deal with that?"

"I don't know," the mayor said, "but the new levee repairs show that we have community backing to do what's needed locally."

I asked how much the current repairs will cost, and who will pay for them.

With help from the Army Corps of Engineers, the town of Hamilton rushed to repair damaged levees between flood events in 2021.

"Rough estimate's half a million dollars."

With a bit of an involuntary exhalation that came out louder than I intended, I commented that I, too, live in a small northwestern town, one with four times Hamilton's modest population but with a local government that still doesn't have two dimes to rub together. Half a million? For a levee that—even repaired—will be no match for the recent flood? The price seemed way high to me, given the work I had seen happening, but even a modest invoice would, I assumed, be problematic in a small and repeatedly flooded town, portable outhouses and all. Hamilton's annual budget, $300,000, must surely be a struggle to raise even in the best of times. "Where will the money come from?"

"We hope to get funding assistance." The Corps was already helping, and if the past was any guide, Hamilton would find the money.

I later read that the campaign for mayor had been laced with widespread though false rumors that the "relocation" would be mandatory and people's homes condemned. I also later learned that efforts to provide resettlement opportunities for Hamilton's repeatedly flooded families continued by Forterra and its local supporters.

At Hamilton I saw that any calculus making sense in terms of safety, economics, climate, and hydrology is made more complex by people's attachment to place and home. I also learned that, regarding options for the future, misinformation is always a part of the mix. Simply getting the story straight can be hard to do. Making the right choice can be harder yet.

FIGHT OR FLIGHT

People have long understood the logic of moving from homes that will inevitably be flooded to homes what won't. But logic is only logic, and a lot of people don't pay much attention to it. The typical response after a flood is to dig out, hose down, repair, rebuild, and wait for the next flood. I've felt all this, myself, after being swamped right up to the doorstep of the cabin where I was living back during Hurricane Agnes.

Leading the list of those who linger in flood zones are people who simply cannot afford to move. As Terri Straka of Rosewood, South Carolina, said after being flooded, "Buy-outs are not going to work for us because of a lack of affordable housing."[12] A lot of people find themselves trapped by the least-expensive real estate, though it's low-priced for good reasons.

Granted: in immediate cash terms, moving to higher ground often does cost more than staying put. But in immediate cash terms, not going to the doctor is cheaper than going, even when you feel a stabbing pain in your chest. Of course, a lot of people simply don't have the money to move or, for that matter, to go to the doctor. One might also argue that those who can't afford to move cannot afford to stay, either. Affordability raises a suite of entangled issues that track all over the map personally, economically, and politically. Since everyone ends up paying for at least some of the costs of flooding, and since many of us recognize at least some degree of obligation to help people in need, aid of some kind is understandably desirable if not essential for those who cannot afford to make the move.

However, for many flood-zone residents, and especially second-home owners, the cost of moving is not the driving issue. Instead, the draw of the danger zone is based on the appeal of waterfronts along rivers, streams, and oceans. The lure of property with a view to the current or to the waves, swept by a breeze tinged with the chill of water, and sweetened by invitations to take a dip on a summer day or to cast a line right there in the backyard—all this pulls on heartstrings and heritage, from bare-bones to trophy-home clientele. People love water, provided there's not too much of it. Any logic, enumeration, or narrative that argues against a waterfront home encounters these timeless seductions, making it tempting to believe that the odds will be with you. "We'll never see that again in our lifetimes," is a postflood refrain I've heard from many who choose not to move in the face of danger, even after being dragged through the mud.

Floodplain management professional David Fowler has noticed that even among people who appreciate the chance to escape risk, and who negotiate the physical and economic challenges confronting them, sadness endures about the community being lost. He recognized, "It's hard to overstate the difficulties that will come with the needed relocations that the changing climate will bring."[13]

Nicholas Pinter, who has studied these matters through the University of California's Center for Watershed Sciences, acknowledged the "social inertia" to remain in flood-damaged homes. He credited it to "the intense sense of place that people have."[14] Street talking a similar idea, a victim of the 2021 Hurricane Ida Flood said, "We rebuild and keep going; that's what we do."[15] Timeless grit and determination seem fundamental to human nature, and in many ways have served us well. But in other ways, not so much.

Like the rest of us, floodplain residents hold tight not only to familiar homes but traditional qualities of places. In the New Orleans of Mardi Gras, jazz, and zesty food, residents loyally cling to the world's only major city centered in a great river delta, which we all know is a bull's eye in the path of America's largest floods, going back to French and Spanish colonists. Many people don't want to move, and even when they've had enough, and get sick of the problems, and do want to go, ties to the familiar bind tenaciously, whether we want them to or not. As John Steinbeck wrote in *East of Eden*, "It is a hard thing to leave a deeply routined life, even if you hate it."

Politically perceptive if not visionary on this issue, Carlos Curbelo, Republican representative from Florida, described a new low-bar for working together in the age of political acrimony when he described flood relocation as "a bipartisan issue in the sense that no party really wants to talk about it."[16]

The age-old stigma over "retreat" from floodplains found contemporary voice with the editor of the *Daily Comet* in Thibodaux, Louisiana, after flooding in 2015. With a keen sense of rhetorical rhythm, though stopping short of solving any problems, his editorial addressed the difficulty of moving from flood zones, repeating six times a catchy refrain: "Move where?"[17] Replacement real estate can, no doubt, be hard to find, especially for low-income victims and those encountering racial discrimination. But remember: 93 percent of America is not floodplain, and a lot of that land is available for new homes. One million acres are being built up every year, presumably offering at least some options for those fleeing floods.[18]

A key dynamic surfacing in people's question, should I stay or should I go? has been termed the "forgotten hazard paradox." Looking into this, geographer Craig Colten of Louisiana State University explained that people have a predictable tendency to forget—even within a few months—a disaster's pains and lessons, real and repeated as they might be.[19] Mayor Richard Lang of Modesto, California, confronted this reality of illusions head-on after a 1997 flood. "When you see a mobile home park float past, you have to think there is a better long-term answer. But I can guarantee that people won't be thinking about that when it's 104 degrees and dry as a bone."[20]

On top of this amnesia effect, the resolve to change is further dampened by what economists call the "status quo bias." We keep doing what we *were* doing. It takes a formidable obstruction on the tracks to derail us, or perhaps an abrupt end of those tracks altogether. But in most cases, memories fade, especially the ones we don't want to remember. People looking for affordable homes often have little concern for why the house they buy is affordable.

For whatever reasons, the option of managed retreat, or what some prefer to call "resettlement" or what I call simply "relocation" and "moving to a safe place," can raise a hornets' nest of opposition from homeowners who have invested financially, emotionally, culturally, and always physically in the status quo on floodplains. Many would rather deal with the tired and exhausting consequences of staying put than with the unknown and unpredictable prospects of leaving the old and finding the new. However, the desire and inertia to stay is increasingly challenged by the disruption and the ruin of property when floods rise, not once per epoch or lifetime as they might have done in the past, but once every decade or even less in the age of the climate crisis. That's what's different about the relocation option today.

The reasons to move are clear: to save money, trouble, and lives, and perhaps even to restore a river for those who are so inclined. But elected officials face counterforces—real, fictitious, and varying shades of gray in between—and often respond to relocation options with a screeching application of the brakes. Their intentions may be to sustain tax revenue to local governments, or to avoid any pretense of telling people what to do. They don't want to spend money on something as distant and abstract as the future—especially when the sour consequences of doing nothing will hopefully not be tasted until after their own term in office expires. Plus, they don't want to risk legal challenges, and in that regard, repeating mistakes of the past that are seemingly inevitable and perceived as normal is less politically risky than making mistakes that are vulnerably new and perceived as optional. Yet regarding relocations away from flood hazards, Katharine Mach of the University of Miami and lead author of the Intergovernmental Panel on Climate Change's *Sixth Assessment Report* urged, "The fear of political suicide should not paralyze those in power from studying the how, where, and why of managed retreat."[21]

WELCOMING CHANGE

Not everyone resists change. Some people embrace it. Some, too, opt for "good" change when confronted with "bad" change. Or, as many of us ask ourselves from time to time about all manner of snags we encounter in life, which problem do I want to have?

Sensible criteria for safety would be important to anyone shouldering the responsibilities of a homeowner, especially when involving the lives of loved ones. In these regards, history shows that risking floods is not a good gamble, but moving to a safe place is. Saving our lives and belongings can in no way be construed as a mistake. I'm reminded, here, of General Oliver Smith. After being insulted by critics when in the nick of time he pulled his division of Marines from the path of annihilation in the Korean War

Battle of Chosin Reservoir, he famously responded, "We're not retreating. We're simply attacking in another direction."

Fitting that bill not far south of Hamilton's flood-ravaged homes, residents at a frequently flooded mobile home park along Washington's Puyallup River in the town of Orting had the opportunity to move off the floodplain, and they took it. Unlike some of the residents at Hamilton, they welcomed the chance to go. One of them, Jon Miller, explained, "Every time they forecast a pineapple express, the anxiety level rose."[22]

Within a system as dynamic and uncontrollable as the weather and the pulse of rivers, change is inevitable, and many people will face the choice to either accept repeated flood losses or head for higher ground. Fortunately, the latter option is also one that—beyond gains of personal safety and economic stability—offers a positive side that can allow former neighborhoods to return to being greenways of water, trees, and trails with room for both people and rivers to wander. That side's now evident along regreening rivers where one can stroll the waterfront and then return to a home that's always dry.

FULLY INFORMED AND FACING THE FUTURE

Much as I did at Hamilton, but a whole width of continent away, I walked the streets of a frequently flooded neighborhood in Charlotte, North Carolina. Only, instead of the wreckage and shambles where recovery seemed beyond salvation or even beyond an informed imagination, I found a spacious lawn and shade trees that arched up where houses once stood.

I tracked down a former resident, who agreed to meet me at her former homesite, now marked principally by a robust evergreen magnolia with its shining foliage and ripening red fruits. Carol Thompson's daughter squeezed a bit of time from her busy job as the office manager for a local doctor and, with her adorable five-year-old underwing, drove Carol to meet me at 5129 Dolphin Lane—a number still painted on the curb but with only grass and trees to show for it.

Carol reflected on the repeated floods she and her family had endured and also on her relocation. "It's not a problem I'd wish on anyone. I loved living here." She paused and looked very directly at me. "Listen and tell me what you hear."

I listened. "Nothing," I replied.

"Exactly. It's quiet here. It was a mixed neighborhood, and all the neighbors were good people. Leaving is not easy. For one thing, a house that floods is hard to sell because no one wants it. For another, even though they give you more than the market will pay for a flooded house, government buy-out programs don't give you enough to pay for the replacement home you need. But no one forced us to move, and none of this had anything to do with racial discrimination. We could have raised the house a little higher, and some people did, but the floods will still come. Raising the house solves only part of the problem. There's no escaping it. The floods are acts of nature. You can't stop them."

Carol and I stepped from the street and onto the softness of the grass, and with just a bit of nostalgia reasonably sneaking into her southern voice, she recalled, "The driveway was right here. The house sat there, from this magnolia tree over toward that other one. Back there is where the creek runs." She pointed across 200 feet of mowed grass to the stream corridor, now carrying barely a trickle, though during floods it raged.

"The Army Corps of Engineers channelized the stream back about 1990, trying to get it to flow faster and not flood, but that didn't help." Carol shook her head, looked me again in the eye, and repeated, "That didn't help at all."

Like others, Carol resisted relocating for a long time. "My husband didn't want to move, and I understood. But then he passed away, and then we had a really big flood in 2008. It was truly bad. Some new houses had been built across the creek, and I think that made the flood in my house even worse. My yard this time was like a river. You couldn't hardly stand up in it. I had flood insurance, but nobody else around here did. Eventually I decided to go, and the county bought me out. I miss the old place, but I'm relieved that our new home doesn't flood. Moving is a tough adjustment, but it's worth it to live where you don't dread the sound of rain, where you can sleep at night when the storm comes, where you don't have to move the car from the low spot in the driveway even when it doesn't rain all that hard."

Emerging from this bittersweet twist in her life, Carol said, "I'm glad I relocated. You move on because you have to. It's a peace-of-mind thing."

Today you can come here where my husband and I lived all those years, and you see flowers and trees. It's now a park, and a pretty place that everybody can enjoy, a place where no one has to worry."

Together Carol and I strolled across the lawn. In the distance a man walked his dog, or vice versa, I'm not sure which, but they both seemed to enjoy the open space. Indeed, it was so quiet that we were like a clip from a movie without the sound running. Then we walked into cooling shade beneath crowns of loblolly pines and past red maples, red oaks, a sweetgum, and also a young sycamore, planted somehow as a sapling not long ago but promising to become a great tree there on the reclaimed floodplain. I imagined it growing old, and sycamores fatten to diameters of four, five, six feet, given enough time. Unlike houses, they thrive on recurrent floods, the roots capable of accommodating months of soaking at a time. We returned to her daughter's car.

I wanted to hear more from Carol, but I hate being pushy with people who share their lives with me, and we fell silent, both, I think, picturing the changes—the unavoidable, traumatic changes that had evolved beneath those trees and across that meadow, now so bright green and inviting. I could only imagine the dark drenching nights, the tumult, the roar, the risks, the voices of urgency, impatience, perhaps anger, and then of sadness and loss, the resolutions that must have been painfully articulated there in a relentless downpour of rain and in floodwaters first seeping under the door, then pooling up to your ankles, rising to your knees, chilling your thighs and headed for your waist, telling you to quit moving your stuff and abandon ship. Carol, of course, knew exactly what had happened and what the emotions and changes were, and they drove her to the conclusions that she generously shared with me. I felt privileged to just walk alongside this woman who had made a courageous choice to look forward in her life, and to strive for something better for herself, for her daughter, and perhaps someday for her little granddaughter, waiting in the car and giggling about something with her mom.

"Should others move?" I finally asked, not wanting to let Carol get away without my learning more from her experience.

"If they can, yes. If the economy allows them. But many can't. If the county had more money, it would help. I'm fortunate. We were able to build a new house, just eight minutes from here. Many people can't do that."

Carol Thompson points out the level of floodwaters before she sold her home and property to Mecklenburg County for restoration as parkland that now marks her former neighborhood.

Funding for relocation is clearly needed, and as I would soon learn, more of that help could be mustered if government aid for disasters and repeated flood insurance payouts and relief were directed instead toward helping people move away from danger zones and be done with the problem.

RELOCATION PAYS

Relocation pays. On contract from FEMA, the National Institute of Building Sciences concluded that buy-outs after floods result in net overall dollar savings to the community and—as the economists say—a favorable benefit-cost ratio. In fact, more than favorable; society saves $6 in avoided costs for every $1 spent to acquire and relocate flood-prone buildings before the next disaster strikes.[23] That's a better return than we get from almost any dam ever built, far better than a wildly bullish stock market yields, and way better than the most optimistic prospects for lingering in flood zones.

Remember, too, in spite of all the talk about not wanting to move, Americans, according to the Census Bureau, move eleven times on average over the course of their lives.[24] That's every seven years on average throughout an American life expectancy of seventy-nine years. Census data for 2019 revealed that 31 million people moved—11 percent of the population—in *one year*. People move for any number of reasons: jobs, medical care, cheaper rent, nicer homes. To be with a loved one, to live closer to family, to live farther away from family, to combat boredom, and on the other hand, to increase boredom in a house less troubled, leaving unwanted strife and hardship behind. We move to breathe cleaner air, to pay fewer taxes, to search for the "American dream," however construed. So why not move to avoid the extreme and certain losses and dangers associated with flooding?

One reason to not move is to avoid facing challenges that are new. But historical perspective shows that moving to someplace new and—let's face it—challenging has been the American way ever since the Puritans and Quakers abandoned the comforts of England, such as they were, for the challenges of religious freedom in the New World, such as it was. Or, for that matter, ever since Asian ancestors of today's Native Americans trekked across the Aleutian land bridge in between glacial epochs, or maybe even before that when they piloted oceangoing canoes across tempestuous seas, taking far more risks than today's floodplain refugees could ever dream about. The notion that endearing values of heritage support staying home has been questioned and overcome in vast migratory movements from east to west in America, from rural to urban and back again, and from bad school districts to good ones. Moving seems to be just as much a norm, just as American as staying put. So why not move from places where disasters are inevitable to places where they're not? Especially when the move might only involve a short jog upslope or to the dry side of a street, or to a new home, like Carol's, eight minutes away.

Regarding the spunk to stay in flood zones, to repair, and to risk again, consider, also, that the battle against floods by building dams and levees and by raising buildings higher while the floods also grow even higher will last only as long as there's enough money to do and maintain all those things. Consider also history's lesson that, irrespective of the money paid, those tired solutions aimed at living with flood hazards have often failed, and are likely to fail even more in the years to come. Furthermore, rarely do the people actually staying in the paths of floods pay the full cost of their decision, and often not even a substantial part of that cost. Because American taxpayers who are not flood prone vastly outnumber those who are, people living outside flood zones—rich, poor, and everything in between—collectively pay the most to cover the cost of both flood control and flood relief for the people embedded within flood zones. That may be admirably generous, but how fair is it?

Deeply entrenched in conventional wisdom, local government officials often oppose relocation because they fear diminished real estate tax revenue, which in fact can fall when we convert homes or businesses to public open space on the floodplain. But that narrow accounting fails to consider the costs of being flooded: expensive damage, danger when driving away as water rises, evacuation of people who elect to stay too long, stress on public services, debris removal, communicable disease in the wake of polluted floodwater, default of drowned-out businesses, recovery and relief expenses at the federal level, deficits of the National Flood Insurance program, further efforts to prevent flood damage such as repairing levees, and—perhaps worst of all from the local government's standpoint—the floods' crippling of public facilities including water lines, sewers, storm drains, power lines, bridges, and roads, any one of which might bust the town budget, Hamilton to Houston. Also failing to be counted are the plus-side benefits of moving: flood zones become community open space, and they enhance recreation along rivers, enriching the economic lifeblood of whole towns and regions. All that accounting gets lost on the ledger when it's time to rally after a flood, or when the costs of flooding are externalized to anonymous taxpayers at the local, state, or federal level.

In years past, when floods didn't come so high and often, tax revenues from floodplain properties may in some cases have indeed helped local governments to operate in the black, but frequent larger flood losses now loom as a formidable blotch of red ink. Or, as spelled out in the textbook jargon of *Natural Hazard Mitigation*, "Floodplain properties generally do not generate high tax revenues particularly in light of the high cost to local government of providing services and periodic flood relief."[25]

Hard data indicate that many expectations of continued tax revenue from flood-prone properties are, in fact, deceptively optimistic. Rather, FEMA reported that after a disaster, 40 to 60 percent of flooded small businesses do not reopen, at all. So forget about that tax revenue. The busted businesses become taxpayer liabilities. Collecting its own data—and by definition striving explicitly to support small businesses—the Small Business Administration admitted that 90 percent of flooded businesses failed immediately or within two years of flooding, even with federal aid. Those are grim prospects for communities relying on tax revenue there.[26]

The economics of getting flooded have always been painful if not debilitating, and now we have the greater floods of global warming. The Federal Reserve Bank of San Francisco published warnings in 2019 that climate change could result in an abrupt end of lending to flood-prone communities. They weren't just talking about flooded homeowners, but entire municipalities. Such reluctance may seem to echo the maligned and regrettable "redlining" of mortgages in minority neighborhoods of the past, but in this case the decisions are not based on racial discrimination but on physical realities of flooding. Whole flood-prone communities will ultimately be affected, including posh coastal retreats with trophy-homes tempting fate during hurricanes.[27] With an eye to that trend, economist Benjamin Keys of the University of Pennsylvania found that real estate prices and sales along many of Florida's coastlines had begun to fall as early as 2013.[28] As risks become more evident, the economic trajectory will likely steepen into a crash of market values wherever land floods. But for now, that prospect often gets tucked away in the circular file of denial.

The economics of flooding are grim, and also complicated. Countering the data of Professor Keys with a time frame going back further, an opposite trend might be equally troubling to anyone who looks: since Hurricane Andrew clobbered Florida in 1992, and amid credible projections for sea-level rise, 5 million people moved *to* the Florida coast in a three-decade real estate boom. A perceptive Abrahm Lustgarten of ProPublica reported that subsidies for floodplain occupation have until now "socialized the consequences of high-risk development. But as costs rise—and the insurers quit, and the bankers divest, and the farm subsidies prove too wasteful, and so on—the full weight of responsibility will fall on individual people. And that's when the real migration might begin."[29]

Regarding that "real migration" from flood zones, Seattle real estate broker Glenn Kelman said, "People don't change unless they have to." He let this sink in, then added, "We're at the point where people have to."[30] Many agree, but the migration is not yet happening at scale. Actually doing what we must is different from simply knowing that we have to do it.

Because of floods and other disasters, 3.3 million US adults, or 1.3 percent of that population, were displaced from their homes in 2022.[31] One in six of them never returned, and this ratio made me wonder what it would take to get more people to move away from floods.

MOVING ON

People resist change, but after each flood, some victims have had enough, and the more frequent the floods, the higher the number of victims who want to move. For them, the daunting scenario of again losing everything and again shoveling mud from the kitchen, buying all new appliances, and fighting black mold oozing from every crack in the house clouds the future.[32] Many flood victims want to migrate to safer ground but lack economic choices and are stuck awaiting the next storm with a constant dread of the siren's wail—if they're fortunate enough to have a warning system within earshot.

After attending a community meeting airing pros and cons of relocation, flooded resident Faye Sesler in Hendersonville, Tennessee, said, "I was amazed that so many people didn't want to sell. What I wanted to do was get them and shake them and say 'listen to this—it's going to flood again.' A lot of people are just scared of moving."[33] With 5 feet of water still saturating his fields following weeks of flooding in 2019, farmer Richard Oswald of Rockport, Missouri, announced, "I'm not going to go back to that house. And I'm not going to let anybody else live in that house because I don't think it's going to be safe."[34] In Nashville Jonna Laidlaw and her husband endured floods twenty times, 2001–19, then eagerly took the city's buy-out offer. "Every time it sprinkled, I got terrified," Mrs. Laidlaw said. Her anxiety had grown from intermittent and temporary to full-time and permanent.

The shift to "mitigation" rather than "flood control" has been supported by a stack of government and think-tank reports dating back decades, most notably after 1993 with what analysts called that year's "unprecedented" critique of disaster management. This approach decidedly pointed toward avoiding flood damages rather than trying to stop the floods. But nothing guaranteed that the government or anyone would actually spend adequate money to do that. Remember, the bulk of FEMA funds goes to postflood relief, usually spent to reclaim and repair flooded homes, just so people can get back on their feet. Furthermore, within the mitigation budget, which includes buy-outs and relocations as a subset, 58 percent of the money goes to "public/private facilities," many of which, such as expensive roads and sewer lines, support continued floodplain occupation if not additional building there.

Within the select group of public funds earmarked for mitigation, only 11 percent went to relocation of structures in the 1990s, and within the government's total inventory of flood-related spending, the relocation budget is far less than that.[35] Nonetheless, between 1989 and 2019, 43,000 homeowners had chosen to accept FEMA buy-out offers, move to higher ground, and allow their former properties to be restored as open space.[36] That was an excellent start. However, 3.6 million people are expected to be flooded annually in coming years.[37] Or worse; First Street Foundation data on projected flood risk indicated that only 0.3 percent of flooded landowners managed to move to drier places in the recent two decades.[38]

Rob Moore of the Natural Resources Defense Council calculated that at the rate of buy-outs during the past 30 years, the next 90 years would see 130,000 relocations. Regarding coastal flooding alone, this is a "drop in the bucket" compared to a projected 13.1 million Americans who could see their homes inundated by a 6-foot rise of sea level at century's end.[39]

Skeptics of orderly relocation hastily point out that large urban areas cannot be moved and must be protected by higher levees and higher first-floor elevations. Of course. Pittsburgh, St. Louis, and Sacramento are not going to be moved. Their presence will call for necessary investments—large ones—and the tax base for each of those cities will presumably cover at least the local cost-share for improved security. However, urban areas large and small represent only about 10 percent of America's flood-prone acreage, or less, and many smaller communities, individual homes, and scattered farms will not justify the economic expenses of stopping a flood or protecting investments sitting in the way.

For the vast majority of vulnerable homes and buildings outside densely populated urban areas, moving to a higher location is the most practical, economic, and reliable approach, especially for structures that get flooded again and again. As FEMA's deputy of resilience David Maurstad sensibly said, "We can't just rebuild back in the same spot, especially if we know that it's currently at risk."[40]

The inevitable migration to drier sites is going to happen, so we might as well do it right, with deliberate, planned, and efficient measures. To do that will require strategic action before the next flood occurs. Facing this daunting task, we fortunately have a lot of encouraging experience to draw upon.

RELOCATION WORKS

In 1972, Rapid City, South Dakota, was poised to rebuild where 238 people had tragically perished in a flash flood only months before, but city planner Leonard Swanson took a courageous stand on the controversial issue and convinced the local governing council otherwise, saying, "We should never permit survivors to spend one more suicidal night on Rapid Creek."[41] Later eulogized for his "unmatched vision" in the path-breaking effort, Swanson directed $48 million in federal disaster aid and $16 million of local funds—big money at the time—to move 1,100 families and 157 businesses out of danger zones. In the process he and others established a 6-mile-long local park and tourism-generating greenway as a new and durable highlight of the city.[42]

Through the 1970s other examples demonstrated the necessity and success of relocation. At Soldiers Grove, Wisconsin, the Army Corps planned to dam the Kickapoo River and build levees, together costing triple the value of all buildings to be protected. The dubious plan was rejected, not by the Corps or American taxpayers, which might have been expected given the losing game in play, but by local citizens who recognized a futile endeavor when they saw one. They opted to move higher and be done with floods. Relocations of thirty businesses, twenty-four apartments, ten homes, and three municipal buildings were completed in 1983. Floods in 2007 and 2008 ranked as 1-in-500-year events, proving the Soldiers Grove move beneficial. In what might be regarded as a by-product, the Kickapoo River remains a beautiful, biologically rich, undammed waterway winding through its cliff-studded corridor.[43]

At Prairie du Chien, also in Wisconsin, 128 households relocated in what Senator Gaylord Nelson—revered for sponsoring the original Earth Day—called a "cost-effective and environmentally compatible" solution to flood problems.[44] Into the 1990s, relocation efforts such as these tended to be special cases where local desires, agency cooperation, and resistance to traditional flood-control structures all serendipitously overlapped for political traction.

Then came the Mississippi's Great Flood of 1993. In its immediate wake, the option of avoiding future floods was totally absent on nightly news, broadcast nationwide and predictably queued on widespread suffering, which, of course, was real. "The reporters showed how devastating the river had been," reflected Kevin Coyle, director of the conservation group American Rivers, "but we saw also an important story about what should be done."[45]

Turning frustration into action, Coyle and his outreach director, Randy Showstack, flung a big media net and piqued the interest of popular talk-show host Larry King, who one week later opened his nighttime TV show by asking Coyle, "How can the next disaster be avoided?" American Rivers' message was broadcast to King's million viewers: rivers flood, and the best thing we can do is get out of the way.

This position had been championed for years by the Association of State Floodplain Managers and numerous government and think-tank reports, but Coyle scored a new level of attention. Seizing the moment, Showstack beat his drum to all the media within reach, and the *New York Times, Wall Street Journal, USA Today, Chicago Tribune,* and *Boston Globe* all editorialized or ran features about the need to keep or re-create open space on floodplains and "give the rivers places to spread out," as the *St. Louis Post-Dispatch* reported from deep in the flooded heartland of America.[46]

Thirty years later Coyle reflected, "There was something about that time and place that resonated with people. The Mississippi was the mother of all rivers. The disaster was momentous. The flood showed that nature was more powerful than people could ever be. We and many others mobilized to move the political dial toward reform—something that had been waiting to happen for decades."

Relocation advocates carried the ball to Congress, and in record time the Hazard Mitigation and Relocation Assistance Act of 1993 directed FEMA to prioritize flood-hazard mitigation—activities that address our vulnerability to disasters or reduce damaging effects rather than trying to stop floods from occurring.[47] The act raised the federal share of mitigation measures from 50 to 75 percent and bumped mitigation funds from 10 to 15 percent of disaster expenditures. Though these may not sound like game-changing margins, they tipped what had been a forbidding balance, launching what Nicholas Pinter of the University of California later called "a new era for flood mitigation." The insurance program was amended to require that payouts for "substantially damaged" structures be used for relocation or to raise first-floor elevations to 1 foot or more above the 100-year level, which at that time was not known to be as inadequate as is known today. FEMA typically paid 75 percent of buy-out costs while local or state governments and homeowners seeking their own deliverance shouldered the rest.

With growing impetus for relocation and with federal funds to do it, Valmeyer, Illinois, and Rhineland and Pattonsburg, Missouri, were relocated after the 1993 flood.[48] In St. Charles County, Missouri, at the flood-savaged Mississippi-Missouri confluence, buy-outs earned a 212 percent return on $44 million in state and federal funds.[49]

Altogether, 20,000 people voluntarily relocated and 8,000 homes were moved or rebuilt in the decade after the 1993 flood. This amounted to 8 percent of the homes that had been inundated and the largest voluntary postflood relocation in American history.[50] Federal relocation aid totaled an unprecedented $375 million at the time—a vigorous federal response and the first time the government seemed intent on changing course from the more-costly and less-effective construction of dams and levees.

The age for moving to safer ground seemed finally to have come. However, with time to reflect, Chad Berginnis of the Association of State Floodplain Managers in 2021 credited the 1990s success not so much to lasting reform but to extraordinary FEMA leadership. "Director James

Lee Witt took FEMA's helm in 1993 and supported the buy-out program in a way that we never saw before," Berginnis said. But with more pessimism in his voice he added, "And he took leadership in a way that we've never seen since."

Successful as it was, the preventative approach of moving upslope totaled only 3.6 percent of federal disaster expenditures for the 1993 flood. Far more money was spent on direct payments for relief and replacement of infrastructure that would await the next flood.[51] Yet, in multiple locations throughout the 1990s, thousands more homes, farms, and businesses were moved from threatened locations, proving the viability of relocation at waterfronts such as Fredricksburg, Indiana. In California, Napa and surroundings responded to a predictable structural levee proposal with a gauntlet of community opposition followed by negotiation with an increasingly amenable Army Corps. The plan morphed into a "living river" theme with restored marshes, widened floodplain open space, breached levees allowing for natural overflow on farmland, and relocation.[52] Likewise, in Milwaukee County, Wisconsin, buy-outs reduced the number of flooded homes by two-thirds. The state of New York bought out 500 homes after Hurricane Sandy. All these strides of the past made me curious about how successful buy-out programs came to be, how some of them have endured, and how they can be increased.

SHOWING OTHERS THE WAY

To look further into the program that made possible Carol Thompson's move in Charlotte, I talked with Dave Canaan, Mecklenburg County Storm Water Services director and the region's official floodplain manager in 2021. "After floods in 1995 and 1997 hammered the city," he reflected, "people began to ask, 'What's going on here?' The water reached far beyond what FEMA had mapped as hazard areas. At Storm Water Services we responded with a lot of dredging, riprapping, and removal of vegetation, all without, you might say, a lot of environmental sensitivity. People recognized that as a problem. We saw their point, and wanted to do better. We knew we had to pull together all the players, so started having stakeholder meetings and listening to all points of view." Though the road of discourse always proves bumpy at times, Canaan added, "Most people eventually agreed that the government should quit trying to control the runoff and, instead, let the floodplain flood. This was not the engineering-dominated approach of old."[53]

The stakeholder group embraced the key strategy to permanently eliminate hazards by "getting up and moving out of the flood area," as Canaan recalled. Setting a new standard, his team transformed a "development-oriented public policy into one that recognized the natural patterns of runoff. It's a whole new playbook on how we're managing floodplains. We also knew that the FEMA mapping was inadequate, so we launched our own mapping program." No one, at the time, suspected how extensive the revisions would be.

County hydrologists determined that the height of floods likely to occur once per 100 years could reach 2 feet higher than what FEMA had mapped. Then the stormwater staff conferred with local municipalities and factored in additional anticipated runoff given the zoning and new pavement in the ultimate build-out of the watershed, which was occurring rapidly during the Sunbelt boom. That raised flood forecasts another foot. Recognition of increasing rainfall upped projected levels yet another 2 feet.

At this point the mapping accounted for 5 additional vertical feet of flooding. Compared to FEMA's minimum rules, the county's precautions cut by half the area where construction was allowed on floodplains. New county rules also shrunk the zones of allowable flood-fortified building to 25 percent compared to FEMA's minimum guidelines. The outcome more resembled the federal insurance program's original intent to prevent new development in hazard zones and to escape from—rather than institutionalize—the cycle of flood, suffer, and rebuild.

"We reasoned that with adequate restrictions on new development we could prevent damages from getting worse," Canaan reflected. "Seeing that our approach was working, people already living in hazard areas began to ask, 'What can you do for *us?*' That's when the buy-out program began. Fortunately, only 5 percent of the county is floodplain, so people had other places to go."

Over the next twenty years the buy-out program saw 75 percent participation among residents facing high flood risks, with homeowners taking the money and moving. The county acquired and razed 425 buildings, reestablishing 192 acres of open space, much of it with recreation access that Canaan noted is "extremely popular." Successful buy-outs included repeatedly flooded apartment buildings, sixty houses that had been evacuated twice in two years, and neighborhoods where greenway acquisitions might ultimately stretch downstream 8 miles to the South Carolina line. Overall tax revenues to local governments didn't shrink but grew, because properties above flood level and next to the newly protected greenways gained 20 percent in value. Canaan explained, "People like living next to parks where you can step out the door and walk or bike upstream or down." While summers grew hotter across the South, neighborhoods next to the wooded floodplain remained cooler. "People recognized that green is good."

The buy-out rate dropped in recent years because real estate values skyrocketed, making replacement housing difficult to find. Counteracting some of that, the program transitioned from its jump start with federal grants toward local funds generated through a surcharge on monthly water bills. Hardly noticed by homeowners, these collectively add up to significant cash for buying properties whenever available. Local money enables the county to act immediately after water recedes—when residents want to sell—all in contrast to the self-defeating hiatus of three to five years for receipt of FEMA relocation money.

Land-use restrictions that bar new development on floodplain open space went unchallenged. An ugly antigovernment revolt that one might expect, given modern acrimonies that fill so much of the news, did not materialize. "Residents recognize that the program protects public safety," Canaan explained. "But going further, this is about community values. We ask people, 'Do you like trails? Do you like clean water? Do you like trees? Do you like greenways?' Well, people like them *all*, and eliminating flood hazards is part of enhancing the larger community. We're creating space for recreation. We're cleaning up the water. We're making these places better, and people who live here recognize that."

As of 2020, expenditures of $60 million by Mecklenburg County, the state, and federal government had curbed $25 million in losses—just the beginnings of a savings account that accrues more with each passing storm and that will overtake costs in short order, if not at the next flood. Keys to acceptance, Canaan summarized, are stakeholder participation along with the agency's ability to show that the program pays.[54]

Essential to continued success are having a relocation plan and delivery system in place before the next flood occurs, having local funds that enable quick action whether or not the federal government can pay, and having capable staff on call. Not every community is so fortunate.

The buy-out sums generously reflect preflood values—significantly greater than what the market pays for a damaged house. But the important number to the flood victim is the cost of replacement. In Mecklenburg County the average new home in 2022 ran $300,000, so even a modest replacement typically costs more than what people get when they sell their flooded home.

"We're now exploring possibilities of Habitat for Humanity helping with relocations," project manager David Kroening said as he guided me to half a dozen buy-out locations like Carol Thompson's. Each had been converted to lawn and trees with new wildness blooming where floods had ravaged former lineups of homes or in some cases just a house or two. Kroening noted that buy-outs often occur when older residents become unable to maintain their home and move to assisted living or some other option.

He suggested that the county's buy-out program is now "maturing," with the highest-risk homes acquired, though even after twenty years net gains in floodplain open space continue to inch upward. "Hopefully in another fifteen years we'll have worked ourselves out of a job buying flood-prone properties and in helping those who stay to reduce future damage."

Another one of the South's booming cities, Nashville in 2010 saw its worst flooding in seventy years. In its wake the Nashville Metro Government compiled a stream basin study that opened pathways for the Army Corps to fund 65 percent of buy-out costs. Local government covered much of the rest. The targeted sites had been flooded repeatedly; all were sold voluntarily. While the flood had wreaked $2 billion in damage, a comparable $2 billion were projected as saved in avoided damage through the buy-outs and regulatory efforts as of 2022.[55]

In addition to the more typical floodplain regulations, the Nashville Metro Government requires, as a minimum setback, that new homes be located at least 75 feet from designated floodways. When more than 50 percent of a house's value is lost in a flood, that house must be elevated not 1, but 4 feet above the 100-year-flood level.

To meet federal and local requirements for subsidized federal flood insurance, homes are often elevated to levels presumed to be above that of future floods, though many of these remain vulnerable in an era of increasing flood severity, and they result in continued problems of safety, municipal services, utilities, and public health during and after floods.

Regarding such flood-proofing measures, stormwater director Tom Palko, who has led the Metro program for decades, clarified, "We still prefer to buy out flood properties and convert them to open space. Even when people raise their homes, problems remain with water supply, sewers, evacuation, and all the usual flood hazards including when people try to drive away during flood evacuations," which, he cautioned, people still have to do even with a flood-proofed home.

After fifteen years on the job, Nashville's floodplain manager Roger Lindsey credited success to recurrent threats of flooding, a history of local governments addressing the issue, and good leadership. He drove me to stream after stream draining into the Cumberland River, and along each, rows of ten, twenty, or thirty homes had been converted to public open space. Bike trails, footpaths, playgrounds, and picnic tables enhanced some of the new greenways. At other sites, just a house or two had been removed, leaving a mowed lawn between remaining homes. Some had been elevated several feet or even a full story. A few tattered remains awaited demolition and regreening of the floodplain. Others, slipping under requirements to "flood proof" because they had lost less than 50 percent of their preflood value, had been spruced up, or not, vulnerably awaiting the next flood.

Pointing to an isolated buy-out in process, Lindsey explained, "We approach this task one lot at a time, but over the years, scattered buy-outs transform into integrated greenways and parks." He explained that while the effort is successful by any measure, much remained to be done. "We've bought 430 houses, which tended to be the ones most-prone to damage. Additional high priorities include another 300. At 1,000 relocations we might be able to turn the page and call the program complete." That goal—and actually seeing that the job could be completed in this sprawling southern city—made me feel hopeful in a new and refreshing way about the possibility of relocating flood victims elsewhere.

As we wrapped up our tour of Nashville's reclaimed streamfronts, the veteran floodplain manager dug a bit deeper beyond the numbers, the acres, and the floods, and while we sat stopped at a redlight, he paused in his story, as well, and then said, "Hundreds of times I've sat and listened to people on the other side of my desk tell me about being flooded, and about the hardships they faced. They came to us with a willingness to be done with the problem. I think that our program has made life here better for them and for everyone." Lindsey seemed to be scanning through a whole career helping people move away from floods, and with an unmistakable sense of earnestness evident, he glanced at me just before the light turned to green, and concluded, "The work has been extremely rewarding."

Tulsa, Oklahoma, also broke notable ground. But not without a lot of pain. Like elsewhere, Tulsa had seen a lot of money spent on Army Corps levees and dams, only to see floods worsen. Up through 1970 the city imposed no limits on hazard-zone development. As a result, 15 percent of the population was living on floodplains. People believed they were protected, but they weren't. Efforts to chart a smarter path first included a privately funded plan drawn up by the legendary landscape architect Ian McHarg, who boldly recommended relocation of homes to higher

As the floodplain manager for the metropolitan government of Nashville, Tennessee, Roger Lindsey has aided homeowners with hundreds of relocations from the urban area's most hazardous streamfront flood locations.

Repeatedly flooded homes along Nashville's Cumberland River and tributaries have been converted to municipal parklands, trails, and natural areas.

elevations and establishment of nature-rich greenways along the Arkansas River and tributaries—an expansive, problem-solving vision that the city leadership promptly ignored in a knee-jerk, cannot-do, pro-development response.

Then came floods in 1974, not just in April but also May, June, September, November. Fed up and motivated, voters approved $22 million in flood-related bonds for buy-outs and open space. This ignited a media-saturated counteroffensive dubbed Tulsa's "Floodplain War," and in the 1976 election, development interests dismissed flooding as irrelevant to the city's well-being and ousted advocates of land-use regulation and relocation. But only two weeks later, as if in sentient revenge by the gods of climate, 10 inches of rain fell in three hours and flooded 3,000 structures. A new moratorium on floodplain development was declared. Yet again it was lifted as memories of the flood faded. More decisively, a 1984 flood hydrobombed the city, inundating 7,000 buildings, dwarfing earlier calamities, and tipping the fickle balance of political opinion. The hazards could no longer be denied. Limits on development followed, this time for real, and the city eventually purchased 1,000 homes for demolition and conversion to greenways, all now a hard-earned relocation spanning fifty years of effort.[56]

Addressing both the city's spotty historical narrative and the reluctance of some residents to move from familiar neighborhoods, city engineer Bill Robinson concluded, "Typically, after people are flooded two or three times, they're ready to sell." The city reinforced its zoning beyond minimum federal standards for flood insurance. Breaking its reliance on congressional grants, the reformed local government—as in Charlotte and Nashville—added a fee on utility bills to fund relocation and flood proofing of homes.[57] For its efforts, Tulsa earned top score in FEMA's Community Rating System, making homeowners eligible for a 45 percent reduction in flood insurance rates. Starting in 1990 this nationwide incentive program grew to include 1,520 municipalities. However, the magnitude of Tulsa's problem remained enormous with a backlog of 24,000 structures still flood prone. In 2019 a flood destroyed another 335 homes.[58]

As in Tulsa, Nashville, and Charlotte, most relocations occur one house at a time when people find replacement shelter that suits them. Other efforts, such as Soldiers Grove, attempt to move flooded communities somewhat intact. Lacking in most of those are the reestablishment of commercial businesses, mixed uses, and critical densities that characterize a vibrant urban neighborhood, leaving the replacement homes in gridded new developments that feel distinctly suburban. Also, no matter how much sense they make, relocation programs are limited not just by money, but by human nature, with difficulties that tend to multiply with larger numbers of people and with the near-unanimity required to transplant whole towns or even whole neighborhoods. Rarely does everybody want to go—at least not all at once—and no governmental jurisdiction wants to tell people they have to do it.

Whether whole-town or single-parcel, and whether a spectacular success or modest effort sparing just one family, it's no secret that floodplain buy-outs could fill a textbook on what needs to be done better. Shortcomings include failure to adequately help people find replacement housing—especially difficult for those who encounter racial or other discrimination when real estate shopping. Affecting everyone, FEMA is notorious for paperwork and a maze of rules, however justified in bureaucratic realms, and delays are augmented by requirements at the state level, where many federal programs are administered.

After the relocations occur and flood debris has been trucked to the landfill, the newly vacant lots can create vexing problems of management: lawns to mow, homeless encampments, and presumptions of virtual private ownership by adjacent landowners who reject buy-out offers and remain, all leading to what planner John LaVelle politely called "unauthorized private improvement of sites that were bought as public open space." Trespasses range from badminton courts to doghouses, fences, and rogue "no trespassing" signs nailed to trees in what are now municipal parklands. The problems illuminate the need to acquire contiguous properties, wherever possible, for efficient management.

Addressing this, Roger Lindsey in Nashville noted that his program tries to buy clusters of lots for the advantages that contiguous greenway spaces offer in park maintenance, public safety, and efficiency of infrastructure including water and sewer lines. "Not everybody wants to sell, but with additional flooding, people typically become willing and even eager to go. Time, which brings more floods, tends to be on the side of relocation. As an agency, we're in this for the long term, and will buy flooded homes whenever we can."

Delays alone kill many relocation efforts. Bureaucracies function slowly, but in this case slow really counts because the target audience has been rendered homeless or severely short on shelter and needs a place to live. After Houston's 2017 flood a foggy nimbus of procedures delayed relocation, which enabled deep-pocket speculators to acquire 5,500 properties that could have been converted to floodable open space. Instead, wrecked and moldy houses were nominally spruced up with cheap paint and resold for easy profits to a new generation of flood victims who were understandably seeking affordable housing and willing to deal with complications later.[59] Yet, in the end, all those difficulties can be overcome. Houston and surrounding Harris County bought 3,500 flooded structures between 1985 and 2020, restoring 1,000 acres to open space.[60]

FOR THOSE WHO STAY

Though it's not the focus of this book, measures other than relocation can be helpful in ameliorating if not solving the flood problem.

FEMA requires that with rebuilding under the insurance program, "mitigation" measures must be taken when damage exceeds half or more of a building's preflood value, and first-floor elevations must be raised to 1 foot above the 100-year crest. The higher the building is raised, the more the insurance premiums drop. Flood "proofing" of this type is often the choice of homeowners and can reduce damage from the next flood—not when compared to vacating the floodplain, but compared to the low-bar of no mitigation at all.

At the Association of State Floodplain Managers conference in 2021, Tammie DeVought Blaney of the Association of Structural Movers—whose business, to be clear, is to elevate or move buildings—explained the logic of jacking up homes, dropping a plumb bob from each corner, and blocking up taller foundations beneath them. She itemized compelling benefits even beyond a drier house: continued tax income to local government, staying in the same house at the same location, and conservation of resources by not having to build a new home or wastefully dispose of an old one in landfills.

In Charlotte, floodplain manager David Kroening noted, "Because of the lack of replacement housing, we're finding that to avoid flood damage we need to turn to mitigation measures rather than relying entirely on buy-outs. Raising the elevation of a house does little to permanently or completely eliminate flood hazards or to enhance riverfront greenways, but it helps with our number one goal, which is to keep people safe." Going beyond the minimum FEMA standard in Charlotte, Mecklenburg County requires that structures with greater than 50 percent damage be raised not just to FEMA's 1-foot threshold, but 2 feet above the 100-year flood.

Officials there found the typical cost of elevating homes was $100,000—cheaper than the average new home of $300,000 but still a questionable option when weighed against the uncertainties in staying on flood-prone land. Complications can include the house falling apart, to one degree or another, during the raising or moving process. The required steps up to the elevated first floor do not suit older residents, and seniors are often the cohort involved. Even without glitches, a raised building is no safety shield against swift current or crushing projectiles, and floods are full of those. After the work is done, safety margins remain in doubt, and the doubts increase as the climate warms. Even with a raised house, floods pose difficulties: evacuation, sewage disposal, water supply, and other problems lingering long after the crest subsides, including black mold and infectious disease. In 2022, for example, flesh-eating bacteria that no one wants to even think about flourished in the wake of Hurricane Ian's floodwaters.[61] In short, for those who want to stay, flood "proofing" has its limits. A lot of work and expense in raising homes is often rewarded by barely keeping pace with the growing flood threats. If that.

The Natural Resources Defense Council questioned the efficacy of FEMA's expenditures to help people "mitigate" threats with flood-proofing measures, and in 2017 they calculated that for every $100 FEMA spent helping people rebuild homes on floodplains, it used only $1.72 to help people move out and be done with the problem.[62] That's a 58-to-1 ratio favoring what many experts consider the wrong thing to do. The Council's Rob Moore succinctly summed this up: "A lot more is spent helping people *stay* in harm's way than is being spent to help them move *out* of harm's way."[63]

REPEATED FLOOD DAMAGE

The most compelling case for relocation applies to homes that are flooded again and again. And there are a lot of those.

Here's how bad it can get: owners of a Houston home worth $114,000 filed sixteen insurance claims in six years and collected $806,000 in federal money.[64] In Spring, Texas, owners repaired a humble $42,000 house nineteen times for $913,000 in government funds. It was not bought out because the flood insurance program allowed and supported reconstruction and continued occupation, over and over.[65] FEMA calls these "Severe Repetitive Loss Properties," though the phrase *throwing good money after bad* may come to mind. No one, of course, would tolerate such losses with their own funds.

Among all the repeatedly flooded properties, one in ten received cumulative reimbursements exceeding the home's total value. This money was granted not so people could move and be done with the problem, but so they could stay and be flooded again.[66]

Out of 2.5 million claims through the insurance program as of February 2022, about a third were repeats.[67] "One-in-three," in this case, may describe the most convincing of all reasons for reforming the National Flood Insurance program.

In spite of persistent efforts to address this affront to taxpayers, this failure to solve flood problems, and this setup for repeated suffering, the problem has gotten worse. Lending historical perspective, the inspector general of Homeland Security reported that between 2005 and 2018 the number of repetitive-loss properties (RLPs) doubled.[68] For floods in 2020, RLPs received $22 billion, much of it as taxpayer subsidies. Fourteen states accounted for 84 percent of the repeated payouts—Texas, Louisiana, and Florida topping the list. Giving new meaning to the phrase *red states,* federal dollars for RLPs flow inordinately in red ink to southern and midwestern states that contribute the least to federal tax revenues that cover the RLPs. Only twelve states—California, Massachusetts, and New York included—pay more in federal taxes than they get back, propping up the rest of the country in this failing economic endeavor.[69]

Repeated payouts during just one year equaled the entire budget deficit of the famously insolvent insurance program, yet Congress extended funding sixteen times in three years without addressing this unbandaged arterial hemorrhage.[70] Losing ground, the National Flood Insurance program in 2020 showed three new repetitive-loss properties for every one that was elevated, moved, or otherwise "protected," even nominally, from flood damage.

Wading through these stats, as serious analysts do, prompted Texas A&M University geographer Sam Brody to leave the usual academic jargon in the office and proclaim, "It is like hitting your hand with a hammer over and over again and never moving your hand away."[71]

Back in 2003 Republican Doug Bereuter and Democrat Earl Blumenauer put politics aside and tried to limit repeat payments. Who, really, could argue with them? Under their bipartisan proposal FEMA would generously offer market value to buy a house wherever claims for damage were repeated within ten years. After acquisition, the government would raze the troublesome structure and convert the property to open space. But developers crushed the reform, vividly described in a *Houston Chronicle* investigation.[72] Reformers were again thwarted in 2012.

For an alternate model, consider Quebec, where Canadians uphold their reputation for sensibility with a generous $100,000 limit on aid to rebuild a flooded house—not per flood, as in the US, but per lifetime of the structure. In Quebec you get $100,000 from the government only once for your flood-prone habit.[73]

Adding to the clustered nature of failings here, a lot of flooded homeowners—even with generous payouts—default on their government-insured mortgages. Banks turn those deeds over to the Department of Housing and Urban Development. But rather than converting the wreckage to open space, as the sister agency FEMA might do, HUD sells the houses, many on the red-flagged, repetitive-loss list, without even a whispered disclaimer about flood vulnerability. According to National Public Radio analysis, 100,000 such handoffs, typically to low-income buyers, occurred in 2017–20.[74]

The Association of State Floodplain Managers' long-running campaign to reform such practices calls for the government to buy out the cyclically damaged homes—so people can move to someplace less burdensome to taxpayers, safety workers, municipal services, and of course the owners—and then convert the damaged houses to open space rather than having

HUD resell them in its subsidized cycle of suffering. FEMA reported in 2022 that acquisition of repetitive-loss properties "continues to be a very high priority."[75] But the spreadsheet of FEMA's flood-related expenses shows that buy-outs are an extremely minor item, and one has to ask, high priority compared to what?

Advocates for a solvent Flood Insurance program maintain that repeatedly damaged properties should be the top priority for reform of federal flood-related efforts—something that Congressmen Bereuter and Blumenauer could tell you it's not. Substantial money saved could go far toward advancing meaningful, helpful, and equitable methods of solving our flood-damage problems.

LOW-INCOME PEOPLE CAN GAIN THE MOST

Particularly troubling in relocation efforts are the circumstances surrounding low-income people. These often include Black residents, especially in urban neighborhoods and across lowlands of the South and Midwest, though their problems typify a larger social stratum in distressed communities of many kinds and every race. Economists at the University of Kentucky documented that people with lower incomes are more likely to reside in places with flood risk because they "live in lower-quality housing, and because disasters exacerbate poverty."[76] In 2021 sociologist Lori Peek at the University of Colorado's Natural Hazards Center investigated the challenges faced by many Black people who get flooded and concluded, "These people have the hardest time recovering." Low-income communities often lack professional staff to access federal aid or, as Jesse Keenan of Tulane University put it, "We have counties and municipalities that do not have the institutional capacity to participate in this alphabet list of programs."[77] In 2022 the Government Accountability Office likewise articulated the need to streamline the flood-related programs of thirty federal agencies.[78]

Nationwide, 450,000 low-income, federally subsidized households occupy floodplains, according to a New York University Furman Center estimate in 2017. Not all, but many, of those residents do not live there by choice and would rather go elsewhere. After two floods in two years, Houstonian Sharobin White voiced apprehensions shared by many:

"It's not safe. Everybody gets to panicking when it rains. You can't live like that."[79]

Consider a worst-case scenario but real-life experience. Low-income residents in Centreville, Illinois, managed to get relocation money and moved with high hopes to a new location behind a levee that FEMA deemed safe, as it does wherever levees meet agency engineering standards. But, as we saw in chapter 5, levees come with uncertainties, and in this case severe drainage problems meant that the relocated people got inundated again, this time not just with floodwater, but with backup from combined stormwater and sewage overflow trapped by the levee that was supposed to protect them. The diked-in wastewater slurry proved to be disgustingly worse than the up-and-down flush of floods had been, with little recourse apparent.

"Problems of disadvantaged communities are deep and complex," cautioned Eileen Shader of American Rivers, a nonprofit group seeking solutions to flooding problems of Centreville and other communities. "Local people need to be put in the driver's seat to find what works for them."

Among the hardest hit experiencing inequities, many Native Americans live on the front line of global warming, which, of course, they had little to do with creating. Quillayute tribal members on Washington's Olympic Peninsula, at the village of La Push, suffer a tightening squeeze wedged between a rising sea and rising river. Similarly challenged 40 miles southward, the Quinault Nation is moving its entire village with 300 homes, businesses, and civic buildings. Among 574 recognized tribes in the US, fewer than 50 have enrolled in the National Flood Insurance program.[80] In 2022 the Biden administration pledged $75 million to help three tribes in Alaska and Washington move above encroaching levels of the sea. Villages of other tribes remain at risk.

Some analysts have noted that White communities have benefitted most from federal money to help flood victims. Lawyer Madison Sloan of Texas Appleseed, a racial justice group, argued, "There's quite obviously a massive disparate impact on the basis of race, national origin and ethnicity."[81] Much of this discrepancy owes to a FEMA calculus mimicking, perhaps unavoidably, the insurance industry overall: flooding of expensive homes rings up larger dollar losses, therefore larger payouts. The agency's federal cousin, the Small Business Administration, likewise follows somewhat standard economic protocols addressing risk and was found to approve 52 percent of its commercial applications in zip codes of predominantly White residents but only 28 percent in Black neighborhoods.

To be clear, FEMA and SBA do not even inquire about race, and David Maurstad of FEMA maintained that because of that, "We don't discriminate against individuals." Following rationale uncriticized in the private sector, federal guidelines call for "reasonable assurance" that borrowers will repay their loans. Arguably a safeguard for taxpayers, this policy nonetheless ends up favoring wealthier clients or at least those with their heads above perilous waters of debt. Looking further, critics make the case that Black communities also have received less FEMA money for repairs to roads, bridges, and hospitals owing to the problems that governments of disadvantaged communities have in navigating complex bureaucracies.[82]

Chief among the problems associated with low-income people is the difficulty in buying replacement housing. Even though payments for government buy-outs cover the preflood value of homes being purchased—and thus much more money than the damaged homes are economically worth—sums are rarely enough for replacement housing in safe locations. Funding to help fill the gap is often available through multiple federal agencies, including FEMA, the Army Corps, and the Small Business Administration, plus state agencies and county or local funding sources, which can be among the most effective, with fewer delays. Thus, the amount of government money for replacement housing can often cover much of the relocation cost. However, even a small percentage of that expense remains prohibitive for many flood victims. To counter the challenges of lower-income people, social justice and other groups, including the Association of State Floodplain Managers, support additional federal funding for relocations.

Striving for solutions to this deep pot of problems, the Anthropocene Alliance represents disadvantaged communities, works for reform of federal rules, and aims to ensure that unscrupulous developers don't exploit relocation programs by acquiring flooded homes and then selling them to wealthier customers.[83]

With a fresh and unprecedented commitment to solving inequities, a FEMA spokesman in the Biden administration said the agency was "advancing" work to level the racial playing field and eliminate policies "that perpetuate systemic barriers" to "people of color and/or other underserved groups."[84] Paul Huang in FEMA's Office of Resilience recognized the need for his agency to "put on the lens of equity and community."[85] In 2022 a FEMA spokesperson indicated that a $200 million grant program for flood upgrades on houses would prioritize areas of poverty.[86]

At the global level the specter of increasing floods and sea-level rise looms monumentally greater. Pacific Institute founder Peter Gleick soberly wrote, "The climate crisis is going to produce two classes of refugees: those with the freedom and financial resources to try to flee from growing threats of advance, and those who will be left behind to suffer the consequences in the form of illness, death and destruction."[87] Without improved national strategies for orderly relocation to safer terrain, Gleick's dark prognosis for developing nations will tragically become our own, with disadvantaged people burdened the most.

That we're already on that path shows in statistics indicating that the National Flood Insurance program had fewer low-income policyholders in 2021 than it did in 2011. Still covering only a fraction of the government's payout costs, monthly insurance premiums in the 2020s (before federal reforms, pending as of this writing) were rising 5 to 9 percent per year. People with modest incomes began dropping out, leaving only one-third of FEMA-listed flood-prone homes insured, many of them owned by wealthier people. The program's decline among the disadvantaged is yet one more reason to pursue relocation as an alternative to the damages awaiting floodplain residents whether they're insured or not.

MORE RELOCATION IS NEEDED

After the 1993 Mississippi flood, relocation programs flourished, but then, echoing past patterns, momentum faded from the postflood flurry, raising the question, was the shift toward relocation, General Galloway style, a lasting reform or simply a moment in history? In the late 1990s construction in danger zones resurged, and it has continued.

Kevin Coyle, who promoted the buy-outs decades before, reflected, "After the 1993 crisis passed and the initial momentum was spent, relocation became mired in all the usual bureaucratic forms. Efforts to increase the program confronted complexities including issues of local tax revenue and lobbying by development industries." Delays of up to five years plagued FEMA buy-outs. While procedures dragged on following Houston's 2017 flood, for example, 40 percent of the people who wanted to sell opted instead to collect repair money and spend it where they could be flooded again.[88]

Take the revealing case of Sidney, New York. After the Susquehanna River washed over 400 homes in 2007 for the second time in five years, the town council and flood victims sensibly agreed that the town should acquire an upland farm and make subdivided lots available for relocation. Diminished tax revenue and community disintegration would be avoided by resettling people within city limits. Officials predicted the effort would take two or three years. Then came delay after delay. Extending water and sewer lines cost an unexpected $4 million. People got tired waiting and took disaster money to rebuild, or sold to speculators, or simply fled, leaving a mess behind while the tax base indeed dwindled. By 2018 good intentions had come to only a few buy-outs instead of a new neighborhood on the hill and a public greenway along the magnificent Susquehanna.[89] The tragic failure could have been avoided if only buy-out approvals had been expedited.

Rob Moore of the Natural Resources Defense Council lamented that buy-out plans get bounced "from local to state to federal agencies and back again, not once but three or four times before a buy-out happens, and that means a delay of up to five years, which is totally unacceptable." Arcane to ordinary people, complications can hinge on whether or not relocation funds are available through FEMA's Uniform Relocation Assistance program or through some other approach, each having its own tangle of requirements. When I asked Nashville floodplain manager Roger Lindsey for a single recommendation to improve acquisition programs, he unhesitatingly answered, "Reducing the time it takes the federal government to approve the buy-out of a flooded home."

Digging deep for a key factor underlying relocation woes, I found author Naomi Klein's "shock doctrine." During and immediately after a crisis, people panic, grasp for recovery, and choose familiar paths even if they fail to solve the problems and even if they open conduits to rogue players having only profits in mind. In unexpected alliance, the politically progressive Klein quoted right-wing economist Milton Friedman, who back in 1982 could have been writing about today's floods when he penned, "Only a crisis—actual or perceived—produces real change. When that crisis occurs, the actions that are taken depend on the ideas that are lying around." People working for change need "to develop alternatives to existing policies" and to keep those alternatives "alive and available for when opportunities arise."[90] A lifetime before Klein's warning, and following the 1938 Los Angeles flood, the esteemed hydrologist Luna Leopold recognized the same lesson. "The willingness of local citizens and flooded-out residents to move to higher, safer ground was frustrated by failure to have ready a plan of relocation."[91] Across the generations the message has been clear: if we fail to take corrective action quickly after a flood occurs, we'll continue to muddle in the same problematic way. To take action quickly, we need to have a good plan in place ahead of time.

But doing that isn't easy. On the coast of Oregon, professional planner Crystal Shoji also served as mayor of Coos Bay. Shouldering multiple responsibilities, she grappled with the preparation issue while facing an ominous quadruple crown of disaster management: flooding rivers, rising sea levels, crushing tsunamis, and earthquakes of magnitude 9 on the horizon, all forecast sooner or later for her economically challenged town. "We need to address the threats of flooding," she recognized, "and the outcome could be positive, shifting development to higher elevations and converting waterfronts to open space. With proper foresight, the result could be both safer and also beautiful. But when you talk about this, people's eyes glaze over. They don't want to hear about it. We lack understanding and a positive vision. We look at how to survive—just to survive—and not at the bigger picture of maintaining functioning communities into the future."[92]

While most towns and cities typically lack the needed preparations for relocation, a lot of flood victims do, in fact, want to move after a disaster. FEMA typically receives requests for several times the amount of relocation funding it has available, leading Michael Grimm, agency director of risk reduction, to say, "There's a clear need there."[93] But where's the money, compared to funds for relief and reconstruction?

Seemingly on the right track in 2021, FEMA adopted new resolve to prevent disasters rather than only respond to them, and Congress funded the agency's Building Resilient Infrastructure and Communities program with $1 billion. But the money was allocated to all mitigating measures intended not only to help people move away from floods and be done with them, but also to control the rise of water through levees and coastal floodwalls, uncertain as their protection may be, as well as for public works demanding mega-dollars for sometimes indiscernible impact in actually eliminating the problem: relocating roads, rebuilding bridges, and upgrading costly infrastructure to accommodate floods, all while people remain in danger zones. Relocation to safe ground faces stiff competition, even within its own line item in the budget. Chad Berginnis of the Association of State Floodplain Managers soberly summarized the problem: "The money is going to big projects, and buy-outs hardly have a prayer in the competition."[94]

In a turnaround noticed by almost no one, FEMA priorities for relocation had quietly shifted in 2014. Up until then, the Army Corps had long been the agency for major flood-control infrastructure, leaving FEMA to cover smaller and arguably more essential homeowner and community work in buy-outs, flood proofing, and house-by-house mitigation for damage reduction. But at this critical juncture FEMA revised its policy addressing the scale of flood-related projects it would fund, opening the door to big-ticket items such as levees, which devour huge bites of budget. The Association of State Floodplain Managers had its ear to the ground and opposed this obscure but bureaucratically seismic policy shift, to no avail. "Ever since," reflected Berginnis, "FEMA seems to have de-emphasized buy-outs and turned away from work that truly eliminates flood hazards. This is disheartening. The push for more 'infrastructure' today includes the same old structural projects that led to more development in hazard areas ever since the Corps built its first dam and levee." In spite of all the advantages of relocation, the key federal agency responsible for it increasingly turned to big projects in a familiar old pattern: build something rather than avoid the hazards; promote more tolerance of the problem rather than eliminate it.

All this may seem ironic, given a prior half-century of hard-earned reform going back to Rita Barron fighting an Army Corps levee in Boston and to the Corps bending away from counterproductive cycles of unconditionally fortifying floodplains for more development, which then get flooded again with increasing damage followed by further fortification in an escalating but futile war against rain, rivers, and nature. But now, with trendy spins about "mitigation" and "resilience" in the face of climate change, the infrastructure ball has been painted green and picked up by the agency once entrusted instead with steering construction away from floodplains and with relocating people away from danger. "You might say that FEMA has become like a 'junior Army Corps' of earlier times," Berginnis concluded.

Furthermore, the impulse to stay and rebuild remained pervasive. It is, after all, engrained in human nature. No agency or person wants to appear insensitive to human suffering, especially following a tragedy, and so the case to resolve flood problems through relocation and avoidance too often remains unheard, just the way it was until Kevin Coyle of American Rivers raised a warning flag and pointed toward a better path after the Mississippi flood in 1993. While earning credit for compassion, which indeed may have been the most important message at the time, FEMA Coordinating Officer Brett Howard seemed to support the goal of remaining vulnerable to future floods—no matter how expensive and how long it takes—when he visited the aftermath of Kentucky's deluge in 2022 and commented, "It's going to take some time to rebuild this. What Mother Nature can do in minutes will take mankind years to rebuild."[95]

FEMA grants can cover up to 75 percent of costs to move people away from flood hazards, but the program purchased only 6,000 homes between 2000 and 2018, hardly denting this problem's armor.[96] The critique extends to farmland. Federal subsidies for flooded farmland after the Mississippi's 1993 deluge ran $20 per acre but swelled to $80 after the 2019 flood.[97] Though explicitly striving to cut public payments for flood damage ever since the original National Flood Insurance legislation was passed, we're clearly losing the battle.

Since early 1993, 43,000 homes have been relocated from floodplains.[98] That's good, but hundreds of thousands remain as priorities. Like the fight against global warming, what we've done so far has only opened the door to what's necessary in facing the challenges of tomorrow.

A NEW VISION OF FLOODS AND THE AMERICAN LANDSCAPE

At this pivotal moment in history, the fate of homes and other investments already located on floodplains could go any number of ways. People aspiring to escape from dangers may be motivated to sell their homes quickly before tomorrow's curve toward greater damage becomes widely recognized in business circles and, more explicitly, in real estate values. The fact that flood risks depress real estate values will be acknowledged sooner or later. When asked in 2022 how the eventual rise of flood insurance premiums will affect the housing market, Florida Realtor Debbe Wibberg answered, "We won't have one."[99]

Then, too, buyers who are not forewarned about flooding may increasingly be duped into investing in lowland property when it looks good—summer, autumn, and dry years—and perhaps at low prices as real estate agents hustle to unload whatever they can, even as storm clouds of global warming darken. History has clearly shown that people are easily fooled, especially when flood disclosures are not required or evident.

Then, too, relocation from flood zones will likely grow as a clear choice of many who become tired of the mess, wary of the hazard, and impoverished by recurring disasters. In the best scenario, the move to higher ground will be made easier with the help of public programs, especially for low-income people, but even with financial assistance, transitions remain difficult. Lacking adequate help, residents now living on floodplains may become increasingly entrenched as they encounter harder times, insidious entrapment with diminishing values, and little ability to move.

The vexing question will be asked, how long will public programs that support staying put in flood zones be sustainable, both economically and politically? As journalist Jeff Goodell more-bluntly noted about coastal flooding, "There has to be a limit to how long the public will pay for protecting beachfront property when we all know it is going underwater anyway."[100]

No matter how much we spend on dams, levees, disaster relief, insurance subsidies, and flood proofing, the floods of the future mean that we'll fail to protect many of the homes and communities that are now located on floodplains. No one has crafted a coherent national plan to address the inevitable crisis. Brookings Institution analysts reported that recent FEMA budgeting for immediate needs of flood relief and recovery outspent by six to one preventative measures such as open-space protection, relocation, and mitigation of all types.[101] That's a lot of money to spend just kicking the can of flood damage down the road.

Given the projections for higher floods, a sensible goal might be for orderly relocation away from the 500-year floodplain—if not from a larger area—in all but the most urbanized areas where levees, at great expense, will be reinforced and maintained.

When the twin tasks of open-space protection on floodplains and relocation of existing development to safer ground are approached in tandem, a remarkable synthesis not only eliminates flood damage, but also restores vital natural cycles benefitting all the life that's associated with rivers, and no other plan for coping with flood hazards can do that. Those natural cycles provide for people in generous and irreplaceable ways at urban, suburban, farm, rural, and wild areas. So let's next consider how the flooding-danger zones of today can become enticing greenways of tomorrow.

10 Greenways

RETURN TO THE RIVER

Imagine, for a moment, rivers with riffles and rapids, deep cold pools, sandy beaches, cobblestone bars, interconnected wetlands, and shading forests where people can walk, bike, swim, canoe, kayak, drive to the edge, watch the flow, see the fish jump and the deer drink, or otherwise appreciate in our own ways the natural world in its most vital form: the passage of water through a landscape it nourishes, all without the threats and costs of flooding to our homes and businesses.

Floods don't have to be bad. Rising water along riverfront greenways is an event that can fill our emotions, not with fear and regret, but with awe at the power, purpose, and beauty of the natural world. Water is good, and a lot of water is very good, provided we're not standing in the way.

So far we've mainly looked at the problems that have come with our historic use of floodplains, and at our strategies for dealing with those difficulties. Let's look, now, at the other side of that coin and see how floodplains have been protected in ways that enhance the places where we live, work, travel, and enjoy leisure time, leaving rivers and floods to do what they've done since Day One on Earth.

In chapter 3 we considered the values of reserving floodplains for floods. Remember, these are the best places for wildlife. Fish need not just water, but the habitat alongside the water. Forests there act as spongy buffers that delay runoff, absorbing it as it seeps from terrain up above, in the process filtering out pollution and suffocating silt before it migrates into rivers and degrades drinking supplies. Open-space riverfronts do all this for free, without costly treatment of water. When rivers overflow, runoff percolates into soil and restores underground supplies that are essential for ecosystem, city, farm, and personal use—no dams, canals, pumping plants, or bottled water needed. Riparian trees keep streams and whole communities cool. Trees nourished by floodwaters grow tall and sequester carbon, forming the front line of defense against global warming.

The rivers' shorelines are prime for recreation where we can touch the water's edge and where blue-green corridors sourced at higher elevations in wildlands curve down past neighborhoods below. Walking, biking, paddling, living with wildlife, and other river-centric pursuits support not just people and towns, but entire recreational economies.[1]

Picturing all this requires little imagination because pioneers of the movement to establish public greenways along streams have succeeded at dozens of waterways. Blossoming efforts continue along hundreds more. We can see them, Seattle to Baltimore. At a larger scale, creating or restoring ecosystems up and down the lengths of our streams has been a key to protecting and restoring nature's connectivity across landscapes.[2] Rivers and floodplains, which necessarily involve floods, form the veins and arteries of the planet, and their management as greenways benefits people, communities, and whole systems of life.

PATHS TO GREENWAY SUCCESS

Because it can affect vast acreage, the most important tool for maintaining riverfront open space—along with the life-affirming hydrologic connections on floodplains—is zoning by local governments. As we saw in chapter 7, zoning can ban some or all development from the 7 percent of America that includes the path of high water and, instead, encourage the building of homes and businesses on suitable terrain elsewhere.

Zoning is essential; however, the most reliable and complete means of sustaining or restoring natural floodplains—the sure thing—is for land trusts or government to buy flood-prone land, thereby transferring full control of its fate to organizations or agencies dedicated to open-space protection. Nationwide groups sharing this mission include the Trust for Public Land, The Conservation Fund, The Nature Conservancy, and others, but more important, regional conservancies and community land trusts acquire streamfronts as priority properties all across the country. The Southbury Land Trust, for example, bought a scenic farm that would otherwise have been developed along the flooding edge of Connecticut's diminutive idyll, the Pomperaug River. The American River Conservancy in California has protected 28,000 acres of flood-prone and adjacent land where the quintessential river of the Sierra Nevada nears Sacramento. Similar groups care for hundreds of streams. Public agencies buy even more flood-prone land. The metro government's buy-out program for flooded houses in Nashville has contributed mightily to 70 miles of greenway trails along the Cumberland River and tributaries. Short of full acquisition, the purchase of open-space easements can allow suitable forms of farming and other flood-compatible uses to continue while safeguarding nature.

Working in this vein, Western Rivers Conservancy founder Phil Wallin observed a remarkable transition in American thought and public policy and succinctly wrote, "Floods have become a reason to restore rivers rather than to dam them."[3] His organization, now under leadership of co-founder Sue Doroff, has elevated the greenway task to an art form, acquiring riverfront property and then passing it on to the Forest Service, state park departments, Indian tribes, or local land trusts for perpetual protection, safeguarding 200,000 acres along 400 river miles and counting. With 400 streams identified as priorities in eleven western states, much remains to be done.[4]

Probing economic factors at play, researchers at the University of Bristol teamed with The Nature Conservancy and found it's cheaper for government to buy floodplains and protect them for recreation and wildlife than to let them be developed and incur the public and private costs of flood damage. Not just cheaper, protection is a dollars-and-cents slam dunk: the researchers found that the cost of acquiring the most jeopardized one-third of floodplain acreage totals only one-fifth the expected cost of flood damage.[5] Buying floodplains and safeguarding them as greenways is the way to go, if only we can figure out how to do this at a meaningful scale.

Well, we *have* figured that out, with a variety of avenues possible. Underwriting many floodplain acquisitions, Congress in 1965 established the Land and Water Conservation Fund to increase America's open-space estate. This engine of public land expansion has been funded by small portions of government receipts from offshore oil and gas drilling. These are dispersed to federal, state, and local agencies for buying recreation and conservation properties—a minor payback for the fossil fuel extraction that now needs to stop in the postcarbon era of climate correction. Over the years much of the money was surreptitiously diverted by Congress from the fund's intended purpose and funneled to unrelated outlets. Establishing overdue accountability, Congress in 2020 mandated the Fund's intended open-space and recreation spending at $900 million per year until federal leasing for fossil fuel drilling ends, as it must, and other funding sources surface. This rescue of the Land and Water Conservation Fund marked a rare triumph of contemporary bipartisanship, indicating that the values of recreation, parks, and greenways cross ideological lines otherwise riddled with political landmines and craters.

Another new federal commitment, the "30 by 30" proclamation of the Biden administration in 2021 aimed to protect 30 percent of America's landscape for nature by the year 2030. This would increase safeguarded acreage from a current 12 percent within national parks, designated wilderness, and national wildlife refuges. Meanwhile, President Biden's hard-fought Infrastructure Investment and Jobs Act in 2021 earmarked $440 million for locally led efforts to increase resilience to the warming climate and also to improve access to the outdoors as a key to public health and community well-being—together a good match for open-space preservation of floodplains.

Popularity of riverfront acquisition programs has been even more evident in state and local bond measures. A leader, if not instigator and key strategist, in passing a remarkable ten voter-approved ballot measures delivering multibillion dollar investments in California open space and parks, Jerry Meral, former director of the Planning and Conservation League in California observed, "Voters repeatedly indicate that they support funding to secure more open space, especially along rivers."

Beyond protection of our remaining natural floodways, relocations of flood-prone homes and farms—and the conversion of those lands to greenways—is a movement driven not only by the "push" of flood damage threatening property and life, but also by the "pull" of healthier river systems and waterfront communities. Rather than sites of repeated losses, floodplains become magnificent public assets when people relocate to higher ground and allow for whole lineups of damaged structures to be removed and for their riverfronts to be restored as natural corridors.[6]

People cherish these greenways. Consider 2 million walkers, runners, and bikers annually at Rock Creek Park's wooded commons through Washington, DC, and the Charles River's wetland parks edging Boston. At just one town in Colorado's Rocky Mountain foothills, 5,000 people daily exercise in the parks along Boulder Creek. At Sacramento, the American River greenway draws 8 million appreciators annually. Floodplain commons along Maryland's rewilding Patapsco River extend for 32 linear miles of state park upstream from Baltimore, now enhanced by critical dam removals as well as open-space acquisitions. Wilder and more remote, protected bottomlands line Mississippi's Pearl River, as they do at any number of waterways traversing national forests and parks—Allegheny River in Pennsylvania to Yellowstone in Wyoming and westward to the Hoh River's blue-green lifeline linking Olympic National Park's glaciers to the ocean and giving salmon a fighting chance to reenact their life cycles, migrating from headwaters to sea and back again.

Working for the conservation group American Rivers, floodplain advocate Eileen Shader said, "A lot of people don't realize the values of riverfront natural areas until they see them. Bringing that sight into focus means re-envisioning our floodplains and converting vulnerable investments into greenways that endure."

Building this case three decades ago, the National Park Service Rivers, Trails, and Conservation Assistance Program compiled evidence—still valid, relevant, and timely—documenting how local economies benefit from greenways. They were found to enhance property assets for adjacent land owners, to boost tax revenue when nearby real estate values escalate, to aid commercial and recreational opportunities such as kayak and bike rentals, to attract new employers to towns whose reputations brighten with waterfront parks, and most important, to cut costs of flood damage.[7]

Riparian forests line the banks of Rock Creek through the heart of Washington, DC, where hundreds of thousands of people walk, run, bike, and enjoy streamside parklands that have been spared from development.

This low-key National Park Service program has helped hundreds of communities transform degraded riverfronts into vibrant recreational corridors. Impressive benefits challenge fears that removing development from floodplains strikes a painful blow through reduced property-tax income. Rather, riverfront greenways typically increase tax revenue.[8]

American history is full of cautionary tales about what happens to the majority of our rivers that are not protected with their greenway values intact. Here's just one outcome that could have been different: in 1930 the noted landscape architect Frederick Law Olmsted Jr. drafted an ingenious plan to link 440 miles of Los Angeles floodplains into an open space and recreational network accommodating rivers, floods, and people in what was destined to become the nation's third-largest metropolis.[9] Unfortunately the city leadership ignored the master designer's advice, and LA became the ultimate freeway city, in the process transforming the San Gabriel and Los Angeles Rivers from the region's most complex and fruitful natural systems into humankind's simplest: ditches lined with concrete spread so wide that city bus drivers have used them as training grounds for learning how to park the ungainly municipal rigs. Modern movements grow to reclaim parts of those streams and their floodplains, though at far greater expense and far less reward than there would have been with the Olmsted plan.[10]

Opportunities have been missed, and nobody says that all floodplain relocations lead directly to criterion greenways. It's often an extended process requiring patience. In North Carolina, for example, Mecklenburg County's buy-outs are impressive but have not always led directly to greenways in the classic sense of contiguous miles laced with bike and foot trails. Sensibly avoiding the use of eminent domain, the program has in some cases produced nodes of greenspace and mowed lots scattered among remaining houses. But those open spaces will grow with further acquisitions, and many will eventually be integrated into a regional network. David Kroening of the county stormwater agency recognized, "Greenways are valuable public assets, and you don't have to succeed one hundred percent to have a program that's beneficial to all."

FROM CALAMITY TO AMENITY AND BEYOND

Once a repulsive alley of pollution-spewing industries, the Boise River through Idaho's capital was transformed into a greenway of magnificent waterfront, irresistible for walking, bicycling, and cool recreational escapes, altogether an exhilarating showcase of the new urban West that draws visitors, residents, and new businesses to the city.

Unusually progressive in 1964 and continuing so today in one of America's ultimate conservative states, the greenway plan grew incrementally through fifty years of diligence in public acquisition of streamfronts. At this semiarid location up against the Rocky Mountains, budding riparian forests now enliven shorelines that once again fit the city's name, derived from the French *boisé,* meaning wooded. "This greenway has been wonderful for everybody who lives and works here," said Hugh Harper, who pioneered the breakthrough and then ascendence of wildlife biology within the federal Bureau of Land Management in the 1950s and later adopted the greenway as a personal mission during retirement in the 1980s. "Dedicating time and effort to a goal with such broad public support and with such obvious benefits for all people was a real plus at the end of my working years," he reflected as the greenway was expanded to 25 miles down river.

With a dozen municipal parks now wedded to the water's path, commuters and kids spin by on bikes. Walkers stroll together and alone, anglers cast for trout, birders eyeball 150 species of brilliant colors, and workers at adjacent businesses and Boise State University chat over lunch at picnic tables beneath towering umbrellas of cottonwoods. At its heart, 6 riffling miles of river draw a festive 125,000 tubers, kayakers, canoeists, and rafters on scorching summer days—one of the West's busiest recreational streams. Swimmers wade and plunge along downtown's refreshing blue-green ribbon. "This is the best thing about our city," said Mike Medberry, resident author and Idaho essayist laureate who has written about the river's plagues, rebirth, and transformation from lowly drainpipe to reclaimed life-source.[11]

Aiming high on a vastly larger canvas, the path-breaking Galloway report following the 1993 Mississippi flood recommended that "many sections of communities . . . become river-focused parks and recreation areas." Going beyond usual meanings of *amenity,* Galloway-inspired hydrologists calculated that restoration of 13 million wetland acres could keep much of the upper Mississippi within its banks during high runoff.[12] Within a few years of the 1993 flood, farmers voluntarily enrolled 50,000 riverfront acres in federal easements whereby the owners kept their land but were fairly paid to relinquish development rights. They continue farming but yield to flooding whenever the waters rise.[13] Though not having "greenway" status involving restoration of wildlife habitat and public recreation, the flood-prone farmland within these easements will at least not be developed.

Some of this acreage lies along the Minnesota River, a contributing artery to the Mississippi. Examining effects of the 1993 flood there, I worked my way down from the stream's headwaters as the historic crest receded, and I paused to talk with the manager of Big Stone National Wildlife Refuge. Jim Heinecke clarified a truth of subtle but profound simplicity, so evident in every view I saw: the riverfronts are not "flooded farmland, but farmed floodland." In other words, the floods got there first and will continue to rise no matter what we do. Most fundamentally, riverfront land is for water. "We may as well recognize that fact and live by it," Heinecke implored.[14]

Similarly encouraging us to coexist with floods along Oregon's Willamette River, hydrologist Philip Williams, at behest of River Network, calculated how protected open space can reduce peak flows in Portland, where floods in 1996 and 1997 crested within a few sandbags of the city's levee-top. Upstream restoration of 50,000 riparian acres to greenways could cut Portland floods 2 feet—a critical margin in downtown's future fate and in the need to push the button for the evacuation siren.[15] Earlier, representing two different political parties but sharing one goal, Governors Bob Straub and Tom McCall had championed a 1970s greenway plan and created six riverfront parks spaced out along 100 miles, with more to come in a vision of riverfront open-space linkage. But then a farmers' backlash, based on principles of private ownership more than practicalities of floods or farms, halted expansion of the popular parks initiative. In spite of Williams's compelling hydrologic data, River Network was unable to persuade additional farmers to sell land or easements for the incomplete greenway, but hope remains that key tracts will become available as the times and economy change. The floods, after all, will get worse, and farming in flood zones is not getting any better.

Facing lesser challenges, and overcoming them, dozens of greenway efforts nationwide have set floodways aside as open space. California's state floodplain manager Mike Mierzwa noted, "As buffers to floods, greenways and parks are becoming more common because they not only give flood flows a place to go but also provide habitat and recreational opportunities." However, protected riverfronts remain a small fraction of acreage facing flood damage. As the Willamette open-space experience indicates, the age of greenway protection is still young. Much remains to be done.

GREEN INFRASTRUCTURE AND NATURE-BASED SOLUTIONS

The idea of addressing flood problems in ways that use or enhance natural functions of rivers has taken many names other than "greenway" over the years, trending from the 1970s term "nonstructural" flood control, and evolving through the 2000s to include "mitigation," "green infrastructure," and "nature-based solutions." Many of these share a common concept: high water will come, so give it room to flow.[16]

Beyond safeguarding riverfront open space and relocating development to higher ground, "nature-based" approaches include watershed management to naturally retain runoff and even out flood crests, improvement of levees by reconstructing them farther back from riverbanks, highwater bypasses that relieve pressure by funneling floods into side channels where waters once overflowed in the historic or geologic past, swales crafted as percolation parks to delay runoff in cities and suburbs, reclaimed wetlands that sponge up floodwater and filter pollutants, and safeguarded riparian forests that capture carbon by converting it from greenhouse gas to beneficially solid carbon in the lasting biomass of trees and soil.

Going back decades, the Water Resources Act of 1974 formalized what are now termed "nature-based" approaches to floods. As this concept matured, the National Wildlife Federation published a 2014 compendium, *Natural Defenses from Hurricanes and Floods,* laying out a wealth of options, all embracing the axiom that "resilience requires working with nature, rather than against it."[17]

In spite of compelling advantages with nature-friendly approaches, shifting policies have made difficult what in the best of times would be a challenging transition from gray to green infrastructure. The political rubber meets the road not with wordsmithing—regardless of how well intended and articulated in policy or statute—but with congressional authorization of money, and for all the advances, green infrastructure remained chronically underfunded, not just in the Corps' budget, which one might expect of an agency with a century of momentum pouring cement into canyons and waterway gaps, but with other agencies as well.

Brighter prospects surfaced when the forward-thinking 2020 Water Resources Development Act required the Corps to update guidelines for natural infrastructure and to evaluate "nature-based solutions" on equal footing with traditional projects, which seemed like a win for fairness if nothing else. Incentives encouraged farmers to sell or dedicate high-water easements, and agencies were required to factor global warming's sea-level rise into construction plans.[18] The act provided guidance to earnest reformers at the Corps who for years had been confronting conflicting directives, such as requirements to select lowest-cost alternatives in flood control no matter what their green-to-gray index revealed. Already working in this direction, the Corps in 2010 had launched an Engineering with Nature initiative, for "integration of engineering and natural systems," and the 2020 resolution to green up America's water development agenda resonated with that effort.[19]

Leading this work with the Corps, Todd Bridges recognized that the old approach to "control or tame nature with concrete had provided mixed results at best" and that our time promised to be "an exciting era with increased interest in nature-based approaches."[20] He pointed out that the Biden administration's Infrastructure Investment and Jobs Act of 2021 mentioned natural infrastructure seventeen times and authorized $5 billion for flood-risk management projects and $1.8 billion for "multi-purpose" work such as levee setbacks. At the Mississippi's Dogtooth Bend in Illinois, for example, the Corps teamed with the Natural Resource Conservation Service to buy easements and increase flood flow capability across 15,000 lowland acres, effectively reinstating floods rather than trying to stop them from occurring. The 2021 law also authorized $2.4 billion for removal or rehabilitation of old dams, including $800 million for foregone safety improvements. Another $3.4 billion went to FEMA for "nature-based" flood projects.

However, the "structure" part of "green infrastructure" can still dominate. Shades of green and gray are easily blurred, and any painter knows that if you mix enough colors long enough, you eventually end up with gray. More troubling than any resistance that skeptics might have expected from the Corps, FEMA appeared to shift not toward the small-scale, locally based, insurance-related measures that the agency had championed in the post-1993 era, but perversely in the other direction. Touting trendy approaches of flood "mitigation" rather than "control," but seeming to regress in terms of where the money is spent, FEMA in 2021 greenlighted

$50 million for a single levee to protect a new high-tech industrial campus being privately built barely above sea level along San Francisco Bay. Though the levee incorporated a flatter slope on the water side for wetlands, this definitively gray earthworks was respun as a "resilience" solution to flooding, even while a rapidly rising ocean on one side of the levee confronts rapidly rising stormwater on the other side for a one-two punch, all underlain by the most eruptible earthquake fissures in America, the whole mix posing the most fundamental question: would a levee even work?[21] Acknowledging complexity everywhere, Bridges noted, "It's not a matter of green or gray, but of finding the correct combinations." He pointed to a separate San Francisco Bay levee improvement that will incorporate restoration of tidal marshes historically dedicated to industrial salt production.

After many pages in this book, a troubling scenario has emerged: newly packaged efforts to keep floods away from people in the 2020s may not be much different from structural plans for dams and levees in the past. As the FEMA-funded levee attests, it's still politically expedient to spend public money building something rather than eliminating the problem altogether by protecting open space and relocating development into safe zones. FEMA took steps toward positive reforms in the Biden administration, but the agency's 2021 directive to focus on "large projects that protect infrastructure and community systems" pointed as much to traditional flood control as it did to the proven success of relocations from hazard zones.[22] The 2021 infrastructure act allocated $47 billion over several years for resiliency, but, as reported by critic Robert Young, geologist at Western Carolina University, "Most of the funded projects are designed to protect existing infrastructure, in most cases with no demands for the recipients to improve long-term planning for disasters or to change patterns of future flood plain development."[23]

Monitoring this disappointing turn of events, Executive Director Chad Berginnis of the Association of State Floodplain Managers urged that federal funds be prioritized for "steering development away from our most hazardous areas" and that "FEMA funding be used to eliminate many of the severe and repetitive loss structures."[24] He criticized FEMA for its new inclusion, if not its favor, of large-scale projects and also its lopsided emphasis on coastal rather than riverine flood areas, as the ratio of spending appeared to be $474 million along 62,000 miles of seashores to $27 million along 2,900,000 miles of rivers and streams. This imbalance may reflect the dire cataclysm of ocean's rise, but it might also stem from the coasts' preponderance of high-value real estate, principally at the Atlantic and Gulf of Mexico.

While FEMA's policies on buy-outs and relocations had not formally changed for decades following definitive success in 1993, the agency's "program priorities" in the 2021 federal Building Resilient Infrastructure and Communities program were clear: buy-outs were "not a high priority unless the project was going to mitigate a large number of structures benefiting a large portion of the community," according to a FEMA spokesperson.[25]

Reforms have included increased funding for FEMA's Hazard Mitigation Relief Program from 15 to 20 percent of total disaster relief payments going to states having "enhanced hazard mitigation plans."[26] But buy-outs are a relatively small part of three distinct "mitigation" programs including big-ticket infrastructure repairs, all sharing the same pool of funds. FEMA typically receives two to three times the number of mitigation requests it can grant, so relocations of homes to higher ground face intense competition within their budgeted line item of mitigation. Even in states prioritizing buy-outs, 80 percent of disaster expenditures still go to traditional relief, covering losses likely to be repeated in future floods.

It remains to be seen how much "green infrastructure" and "nature-based solutions" will amount to reconfiguring or perhaps just redescribing projects that are fundamentally gray and structure-based.

Yet, the trend overall elevates natural processes in flood management. Optimistic about the movement toward "nature-based solutions," *Floodplains* author and World Wildlife Fund scientist Jeff Opperman noted that "a lot of effective projects for flood-risk management will still need to include structures. A levee setback still needs a levee. There are multiple benefits, to both people and nature, from large-scale projects offering some blend of what's natural and what's managed."[27]

WORKING TOWARD CHANGE

Recognizing the limits and problems of past efforts for flood control through dams and levees, along with the promise of working productively with natural processes rather than against them, Andrew Fahlund of the Water Foundation asked a key question pertinent to our nation's response to floods: "How do we turn this ship?"

With the long view in mind, he found optimism in the coalitions of people affected by flooding and now working together. "For years the discussion was dominated by major players. In the Central Valley of California, for example, that meant the agriculture industry and a handful of cities, which clung to a time-tested agenda: more dams, stronger levees. As the limits of those approaches became apparent, the traditional and entrenched groups struggled not only for public endorsement, but also for the government money that once flowed freely to build flood-control structures. People began to see that multi-benefit projects were the future. Voters in California and elsewhere passed water bonds favoring work to protect the natural environment as well as to serve farms and communities. Eventually people from many persuasions realized that they were better off working together."[28]

John Cain, with the hands-on restoration group River Partners, sought to shift away from the stolid old structural way but admitted the enduring importance of unequivocally putting safety first. "We're not going to argue about safety, and we say that clearly. Then, once a person or group is open for discussion, we talk about 'multiple benefits.' In today's world, people get the idea that big investments in taxpayer dollars need to serve a wide range of needs. These include the water that rivers deliver, the recreation that people seek, the habitat that streams provide, and the greater natural systems that nourish communities in fundamental ways."

At the cutting edge of reform, the Tuolumne River Trust launched California's largest riparian restoration project, carried forward by River Partners to reinstate flood flows after breaking down old levees and planting native trees on 2,200 acres where the Tuolumne meets the San Joaquin River, enshrined in 2022 as California's newest state park.

South of there, the San Joaquin may be an unexpected candidate for restoration and greenways in spite of a history of indiscriminate damming, desiccating withdrawals that reduced flows of a major river to zero, and irrigation return water more fit for hazmat overalls than bathing suits. The river floods infrequently but occasionally bursts over its banks with the increasingly rapid melting of Sierra Nevada snowpack. An impressive flood of 12,000 cubic feet per second in 1986 was dwarfed by 70,000 in 1997, five times the projected 100-year event. Enormous flows followed in 2023. Amping up the alarm significantly, forecasts call for more rain and less snow at high elevations and for flood crests at a shocking five times the 1997 deluge. "That's daunting," noted Cain, "but fortunately the unique characteristics of the San Joaquin support a strategy of river restoration."

Explaining such optimism, Cain pointed out that winter rains are followed by Sierra snowmelt, which continues during the summer growing season, historically preventing most acreage close to the river from being heavily settled or, for that matter, farmed. Floods have overtopped local levees, and future floods will overtop them by far, but in the extreme climatic fluctuations of the global-warming era, the basin more typically faces ominous droughts requiring that overall agricultural production be reduced. "Tomorrow's trends for both floods and droughts support the reduction of marginal farming along the river, and this means that some levees could be dismantled," Cain explained. "They'll be ineffective anyway during the big floods to come. In the wake of all this, we see opportunities for restoration of riparian wetlands, floodplain habitat, and underground aquifers serving a vast area where farmers need replenishment of groundwater far more than they need flood control along the river."

Possibilities likewise emerge in the Pacific Northwest, where a Floodplains by Design program of the Washington Department of Ecology combines efforts of state agencies, local governments, and nonprofit organizations to increase safety and restore rivers for spawning salmon. The initiative seeks to reclaim 15 percent of Puget Sound's historic floodplains—hopefully just a start.

Pioneers of green infrastructure also work in our largest urban areas. Matching the scale of New York, 11,000 rain gardens have been built to catch runoff from streets and rooftops—an image-boosting greening of America's largest city, once the ultimate in gray infrastructure.[29] A model for smaller cities, Charlotte requires parking lot owners to add grass, trees, and bioswales strategically located to delay runoff, some of them connecting directly into Mecklenburg County's network of open space that's incrementally replacing flood-prone homes and whole blighted blocks.

In converting the hazards of floods into the benefits of floodplain greenways, there's no shortage of models that can be followed. Creative solutions can turn the fate of floodplains from corridors of destruction into pathways of life.

11 Living with Rivers

WHAT NEXT?

On a stormy day I watched waters of the Hurricane Agnes Flood rise along tributaries to Pennsylvania's Susquehanna River, and the compelling nature of that flood marked the beginning of a long and fascinating journey. The 1972 deluge spared the cabin where I was living, but only by inches, and neighbors suffered deeply. Seeing the rise, watching its power grow, and shoveling mud in the aftermath, I resolved to never own a house on a floodplain. I was fortunate, not already owning one.

Others faced painful choices. An immediate impulse says to stay and navigate the path of least resistance. By default if not design, we do what's required to reestablish life as we've known it. But when the route ahead appears not only uncertain but also worsening, when the chances of further losses are understood, and when the full accounting of not only risky hazards but also exhausting needs, painful costs, and attractive alternatives come to light, other possibilities float to the surface.

Not yet knowing what all those were, and stranded in my village as the water receded, I tended to emergency needs. I immediately noticed that the flood had deposited clearly identifiable mud and a wrack-line of leaves and trash at the precise level of the crest. It was so high and in places so far back from stream banks that I knew its extent would later be doubted, disbelieved, forgotten. Yet it's important to know our history and that of our rivers—from the details of one flood's rise to the grand scope of our high-water heritage—if we're to have any chance of improving our plight.

The day the road opened, I hustled back to work as the county's environmental planner and, remembering that dirty bathtub-ring that Agnes had left as its calling card, immediately scrambled a map-savvy team on our staff to field check the upper limit of the water while we could still see it, and to pencil in the critical line on topographic maps. For two weeks we field checked and recorded the flood's peak along hundreds of miles of waterways. That charting of Agnes's historic crest became the only floodplain map in our county for some time until FEMA rolled out its official insurance maps, which, of course, took a while.

My boss and the planning commission we worked for recognized the need to act immediately before flood memories slipped away, and we urged local government leaders to adopt or revise zoning to prevent new development on the floodplain. Within two years our planning staff helped all fifty-two local towns and townships enact floodplain ordinances—something we had earlier failed at because elected officials hadn't seen the urgency or believed that the floods of 1889 or 1936 would come again. Agnes showed they would.

My career and life took me in other directions and to varied crises and opportunities on other rivers, but in the coming years planning director Jerry Walls expanded efforts in a buy-out program for people suffering repeated damage. "We found that many of our losses occurred in a few troublesome pockets that were flooded over and over," he recalled in 2021. "We were able to garner local political support and then raise state and federal money to buy dozens of homes and convert riverfront properties to parks." Ultimately Walls's thirty-eight-year tenure as county planning director provided the continuity and commitment necessary to make the relocation program work. Carrying forward, planner John LaVelle said, "We've had 200 houses razed or moved from floodplains, and we keep chipping away at the problem."

Seeing further how public land acquisition can boost the recreation potential of waterfronts at towns along the Susquehanna and thereby

The Susquehanna Greenway Association seeks open-space protection and recreational enhancement along 200 miles of the Susquehanna River where extreme flood damage in 1972 and subsequent years led to efforts for better management of floodplains.

prime the pump of local and regional economies, Walls, undiminished in "retirement," led formation of the Susquehanna Greenway Partnership and poured his substantial energy into a mission soon backed by twenty-two counties spanning half the state—one of the geographically largest such greenway efforts anywhere. "Our goal is to establish public recreation sites and trails, and in the process transform flood liabilities into public assets," Walls explained. "Strategic floodplain management turns our riverfronts into community resources with trails and access to the water. That's good for everyone. The floodplain assets—rather than hazards—become economic drivers that attract people from twelve states for recreation along our local waterways." Ultimately the greenway, in one form or another, will extend from Chesapeake Bay up the Susquehanna and its West Branch to Pine Creek, a state-designated Wild and Scenic River 200 miles from the estuary. From there, a 60-mile waterfront bike trail on an abandoned railroad right-of-way links upstream, and its extension could cross the eastern continental divide and descend New York's Genesee River to Lake Ontario. This epic job of connectivity, cooperation, and imagination is far from complete, but today we can already walk or bike along many miles of greenway once doomed to flood hazards. Dual concepts of minimizing flood damage and enhancing recreational open space have merged into a powerful and positive force for the future. Walls's vision can give us all hope. At the Susquehanna it became clear that committed people can begin to turn around the flooding problems of the past and work for something better.

Similar vision has emerged along many other rivers and come into sharpened focus only vaguely imagined fifty or even twenty years ago. Relocation at Rapid City, South Dakota, and Soldiers Grove, Wisconsin, marked early successes by helping flood victims move to safer terrain while establishing greenways of enduring public value. Mecklenburg County, North Carolina, later rewrote the floodplain restoration playbook, as did others at Tulsa, Nashville, and dozens of communities. The narrative of flooding has become not only one of tragedy, but also one of viably durable solutions. As California floodplain manager Mike Mierzwa wrapped up the new credo, "Floodplains do more for people as places that flood than they do as places for a house or business."

To bracket my personal experience that began with the 1972 Hurricane Agnes Flood, I chased the crests of historic runoff, winter and spring of 2023 in California. Floods in winter, alone, were impressive as rivers such as the Russian rose and fell. Then, monumental accumulations of snow buried the Sierra Nevada in 20, 30, and more feet of snow, all staged for melting as April turned to May. I gazed there in awe at the makings of floodwater, and California had been flooding for the past six months! Springtime snowmelt pushed the American, Stanislaus, Tuolumne, and other streams over their banks while the Kings, Kaweah, and Kern triad of southern Sierra rivers refilled much of what had long been drained for cotton fields in the historic bed of Tulare Lake. Once the largest lake in the West, this former commercial fishery and inland sea, four times the size of Lake Tahoe's famously blue expanse, had been long forgotten, but its rebirth, too big to see across, spoke of the power in rivers that prevail and of the climate that's warming with potential for storms bound to exceed those even in 2023. Floodwaters that recreated a lake unseen for generations reminded me that the hydrologic cycle rules, and that our efforts to slow it, to speed it, or to deny its fickle yet absolute imperatives will, in the end, be futile.

The power and persistence of floods in 2023, not just in California but also the upper Mississippi, Pennsylvania, New York, and Vermont, served as a reminder and a compelling obligation to accelerate, bolster, and propel efforts to make way for even greater flows to come. As climatologist Daniel Swain commented during the crest, "We're nowhere close to a plausible worst-case flood scenario for California."[1] Precipitation for six months totaled an impressive 2 feet of rain across the state, much of which is usually semiarid. But get this: in 1862 roughly *ten* feet fell in forty-two days. Now factor in the spike that global warming delivers. As hydrologist Dettinger wrote in 2013 and as Swain warned ten years later, a really big flood will happen again. It's just a matter of time.

Everything I've learned, from Pennsylvania's Hurricane Agnes to the West Coast's ongoing atmospheric rivers, leads me to question how we might defy our tired and catastrophic history of repetition in flood losses and instead do something better. Now, given both the past and the projections that this book has examined, what will tomorrow's floods bring to America? Consider these outcomes.

Our dams and levees will continue to serve some of their intended purpose of controlling high water but will predictably fall short in the largest floods. Relief payments and insurance will ease the immediate pain but will do little to help anyone escape the aggravating if not tragic cycle of loss, anxiety, debt, and repeated loss. Likewise, incentives and regulations to "flood proof" new construction by raising it 1 foot or even substantially more above expected high-water levels will encourage an even more misleading sense of false security in the face of escalating storms, and will ultimately prove futile.

As inevitable as the landing of a rock thrown off a cliff, the climate will continue to warm, even presuming we end widespread burning of fossil fuels, which promises to be a struggle that we're thus far losing. So, flooding from rivers and streams will increase, and we don't know how much. Anybody counting on currently projected levels is gambling, often with an entire life's estate at risks never imagined. Larger floods loom not in the realm of chance and theory, but in the realm of certainty and fact, or as the Intergovernmental Panel on Climate Change carefully understated in its sixth assessment in 2021, "Climate change is intensifying the water cycle. This brings more intense rainfall and associated flooding."[2]

Short of the long-term task of halting global warming, those at risk can do nothing to prevent floods from occurring and worsening. Our current pattern of privatizing profits from floodplain development while socializing the costs of flood damage will no longer be tenable. So what needs to be done?

First, visionaries of change along with advocates of precaution say that personal accounting for one's self, family, and property is crucial. "In 1994 we stressed that individual citizens have the responsibility to look out for themselves," recalled General Gerald Galloway in 2021. "Today, that message is even more important. It should not be undercut by government programs leading in the wrong direction, but rather guided by programs pointing to the route that we must follow."[3]

This book's search for that route through the kaleidoscopic geography of America and also through the tragic history of flooding recognizes that the most effective solution is the simplest: don't build or buy a house if it's on a floodplain. Consider the threshold of a 500-year deluge and more. No one knows where the upper limit will be, but realize that atmospheric carbon and its climatic imperatives have already increased to their highest levels in 3.6 million years and will continue to get worse even under the most hopeful scenario.

Furthermore, a precautionary approach aimed at keeping people alive and our belongings intact requires this: if you already live in the path of a flood, move out when you can, ideally selling your property for incorporation into a park or waterfront preserve. Of course, not everybody is able to move, but for those who can, simply getting out of the way could solve much of America's flooding problem. To put this in marketing context, buy security before you suffer the loss, and bear the cost of security before the move becomes too expensive. Again: not everybody is able to move, so let's do all we can to help those who want to move but face the greatest challenges.

Individual initiative counts, as floodplain management advocates from Gilbert White on down attest, but its limits are real, and many if not most of society's crises can be sourced, at least to some extent, in the failure of individuals to independently do what their own and also our collective well-being requires. This reality means that our governments have an essential role to play.

Though antigovernment rhetoric often dominates the news and everyday culture around us, especially in rural areas that include the majority of flood-prone land, a reputable 2020 survey revealed that two-thirds of Americans think government is not doing enough to address climate change.[4] So here's a chance to do more.

It makes sense that people will respond when problems affect us personally, and flooding is certainly one of those problems. While the global climate crisis looms, with both causes and effects seemingly intractable to many of us, addressing flood hazards is one constructive way to engage and move forward with practical, doable solutions in every community, city, and state. The time has come to let rivers be rivers, and to recognize that they are forces of life greater than we are.

To collectively do what's needed is unfortunately a personal, practical, and political stretch for many, so let's talk about what should—in all three respects—be easy to do.

First, acreage that still remains as undeveloped floodplain should be zoned as permanent open space. Just leave it the way it is. This status-quo option describes, in the truest sense, an extremely conservative notion: avoid change. Case law has supported this kind of public-interest land-use regulation for a century, and floodplain zoning, to one degree or another, has already been enacted in most municipalities, though it needs to be more effective. Only in this way can we shut off the symbolic "spigot" of increasing flood damages. Accomplishing this task will enable us to then address the corrective job of "mopping up the floor" with relocation of damage-prone investments, establishment of greenways, and creative pursuit of other means to give rivers room to flow.

Here's something else that should be easy to do: disclosure of flood risks. This would be an aid to anybody who contemplates buying or renting a home on a floodplain. Unnecessary as it should be, legislation is needed to make this information readily available from FEMA and also mandated in real estate transactions the same way that deed searches are required to protect investments of buyers.

Looking further, here's an imperative that really counts: our governments should stop subsidizing rebuilding in repeatedly flooded areas. The real estate and building industries that have opposed floodplain restrictions and limitations on rebuilding in hazard zones might do well to recognize that reserving floodplains for flooding requires not that development end, but that it simply migrate to places well-suited for it—to places where the lasting legacy of the developers, themselves, will be one of security, permanence, and gratitude rather than suffering, loss, and blame.

Key to any long-term success, government programs need to transition away from repeated flood relief and insurance payouts and aim instead toward floodplain buy-outs that would end, rather than institutionalize, the costs and the suffering that otherwise are bound to occur again and again. Truly solving the flood problem through buy-outs should no longer be the poor cousin, in appropriations, to other government programs aimed at relief, dams, levees, rebuilding, including even today's popular "green infrastructure."

Regardless of political persuasions, we should be able to transition away from spending billions of dollars shoring up the most imperiled of our dams, levees, and short-sighted infrastructure, much of which will inevitably become ill-fated as global warming and its rising floods confront aging defenses that were inadequate if not ill-conceived even the day they were built. Maintenance and improvement of critical flood-control structures will, of course, be needed to protect densely urbanized areas, but good money should not be thrown after bad where the efforts of the past simply don't work: dams that fill with silt or pose safety risks beyond repair, and levees destined to fail in tomorrow's storms because constituents cannot afford the extraordinary costs of upgrading those structures to fail-safe standards in places where few people are available to pay the enormous price of a waterproof fortress against floods.

Looking more to the bright side of these efforts, transitions to greenways provide the ultimate solutions to flood problems, especially in rural areas and riverfronts where urban, suburban, or even scattered housing developments have not yet occurred. The expense of greenways might better be considered not an expense at all, but rather a high-return investment compared to the private and public cost of repeated flooding, and also compared to the bankrupting flood-control efforts that we've waged for a century with disappointing results.

Today we stand with one foot in the past by trying to control the flow of water during floods and one foot in the future by trying to avoid the problems brought by the next blast of an atmospheric river or the next violent spin of a hurricane. To step forward requires readiness to move in momentous ways when the next big flood arrives, as it surely will. The inspiration and the leverage to prepare for that move can come with each and every flood from this day onward. As Winston Churchill reportedly urged while supporting formation of the United Nations following World War II, "Never let a good crisis go to waste." Only with the required preparation will the repeated crises and failures of the past yield to a better future. Yet many communities are not ready to do what's needed, and neither is the federal government.

I've asked dozens of people engaged with flood work in political leadership, public agencies, and nonprofit organizations why more is not being done to avoid new development on floodplains and to aid in relocation above hazard zones. The answers span the scope of resistance described in earlier chapters of this book, ranging from economic to cultural, political, and personal. Much of the failure boils down to the fact that the shifts we're making are "too often incremental," as Camille Parmesan, a University of Texas ecologist and co-author of the Intergovernmental Panel on Climate Change's Sixth Assessment, said. She noted that our time requires "transformational changes."[5] Fortunately, even transformational changes don't need to be intimidating or forbidding. They can be navigated like other great journeys: one step at a time, which in this case can mean one flood at a time, one town at a time, and one house at a time.

THE POSSIBLE FUTURE

Addressing "natural" hazards of both natural and unnatural origins, a University of Colorado study found that the riskiest acreage of American landscape—variously vulnerable to the ominous sextet of floods, hurricanes, wildfires, tornadoes, droughts, and earthquakes—cumulatively covers 30 percent of the country. This sizable composite presents a daunting challenge affecting not just 30, but 60 percent of buildings—and one might infer people—nationwide. The disproportionate number owes to development gravitating to America's coastlines, with their extreme threats of storms and floods, and to the seismically active West Coast.[6]

But unlike hurricanes, wildfires, tornadoes, droughts, and earthquakes, river flooding occurs on a relatively small and defined acreage, just 7 percent of America, and less than that beyond the Midwest and South with their wide swaths of lowlands. Unlike with the other hazards, we know exactly where floods will occur. For this reason, high water should be the easiest of the six troublesome types of "natural" hazards to address and avoid.

Where people and local governments have the will, they can prevent flood losses and reinstate essential natural workings of the Earth and its waters. The determination to do that may not be apparent, but California river protection veteran Ron Stork retained modest but dogged optimism saying, "When there are no other options available, we'll do the right thing." Earlier chapters of this book argue that we've exhausted other options, only to see the threats and damages of floods mounting higher.

Flooding occurs on explicit and limited tracts of land, yet it's a problem that affects everyone, so the topic of this book might be considered a unifying issue in America today. Let's consider the two political poles, at opposite ends of the public opinion spectrum. Doing what's needed regarding floods appeals to what were once perceived as classic conservative values: minimizing government expense, acknowledging personal responsibility, and in the true spirit of conservatism, avoiding changes that threaten what people cherish—a directive that clearly includes the changes and destruction when floods bury homes in water and mud. Doing what's needed to address flooding also appeals to traditionally liberal pursuits: helping people in need, leveling the playing field among social classes, and protecting the environment. One might regard flooding as a problem to be solved by setting aside our differences and engaging in goals about which sensible people agree: save lives, reduce our damage from floods, and nourish healthy rivers that serve everyone.

Through much of our past, responses to flooding from both our political poles had a great deal to do with combating the perils of nature, building and rebuilding with bluster and resolve, and moving forward with an uncompromising belief in progress, as it was thought to be. Responses in the future might benefit from similar determination, but they'll need healthier doses of imagination, practicality, humility, and ultimately wisdom about our figurative and literal place on Earth. Addressing the flooding problem pushes us further than we've collectively reached before because the high water of the future is driven upward by the ominous warming of the planet's climate. Tomorrow will be a new, different, and more demanding world.

Which brings us, full cycle, back to page one of this book and my own story of record-breaking floodwaters lapping ominously higher and higher on the riverbank where I so precariously lived. As I write this line in 2023, I see the same threats, damages, and losses occurring again in many other places. The story of flooding remains one of victims' suffering along with public expense and repeated damage. But this story can be changed whenever we summon the collective will to look in new ways at where and how we live.

We have no choice but to face the challenges of the future, and how we address them will depend on our ability to alter the failed patterns of the past. Bold programs of federal and state governments need to recognize the enormity of our flooding problems. Each community needs to accomplish its own increment of change even if it occurs one house at a time. Each person needs to avoid the cycle of tragedy and loss when the levee breaks or, more often, when it just happens to rain for two days at a time.

A workable, positive, and even bright future requires that we begin now to prepare for the high water that will come. We can gather wisdom hidden in our history, illuminate it with modern science and the logic of economics, and in the end, face the floods of tomorrow in ways that solve the problems of the past. All of that leads not to anxiety at the sound of rain on the roof, but to finding beauty in the river's rise and peace in knowing that we live within the ways of the natural world.

To do all of that requires a new respect for the Earth in all its timeless patterns and imperatives. Chief among them is water flowing toward the sea in amounts that can, at times, be enormous. The richness and power of that flow make it essential that we not stand in the way.

The age of denial is over. With the passage of one era and the arrival of another, an age of increased flooding is here. It's time to address this problem from the past and to face a future that's not plagued by rivers of destruction but nourished by rivers of life.

Tomorrow will bring greater floods whether we plan for them or not. It's our choice to live vulnerably in their path or to seek higher ground.

Acknowledgments

Many thanks to my wife, Ann Vileisis, for her sharp insight, unwavering support, and wise editing suggestions that helped me throughout. Editor Stacy Eisenstark sensed the importance of this topic from the beginning. Editor Chloe Layman expertly ushered my book through the extensive review and publishing process at the University of California Press, and with great kindness and insight encouraged me ever onward. Lou Doucette skillfully copyedited the final draft.

For fact-checking, quote confirmations, and more, I sent many pages to experts including virtually all the people I interviewed. For reading and review of the full manuscript, deep appreciation goes to David Fowler of the Association of State Floodplain Managers; Jerry Meral, former director of the California Planning and Conservation League and deputy director of the California Department of Water Resources; Jim Palmer; and Becky Schmitz. Phil Garone, Peter Moyle, and Bob Freitag also reviewed the final manuscript and offered encouraging and insightful comments.

For early inspiration to engage in the problems and the promises of flooding, thanks to Jerry Walls, director of the planning agency where I worked for eight years; Chris Brown, former director of American Rivers and of river programs for the National Park Service and US Forest Service; and Dr. Peter Fletcher, professor of forestry, who first introduced me to the importance of floodplains when I studied landscape architecture at Pennsylvania State University.

Sources

Facts, knowledge, and citations here come from both primary and secondary sources: books, articles, news reports, conferences, interviews, photographs, art, films—whatever I could get my hands on. I've included endnotes for critical data when I suspected that readers might question the source or want guidance to further detail. I often selected reliable secondary sources because they are easily found and offer broad journalistic context. Many of my sources also come directly from agency, academic, scientific, and nongovernmental organization reports. Most information is available online and can be found with only nominal difficulty given the authors and titles listed here. Without sacrificing essential details needed for retrieval, I shortened traditional footnote style to save precious space in my narrative, as my book had a specific length limit. I also built my narrative on direct interviews with professional experts and people intimately involved with floods.

Among hundreds of published sources, let me mention here a few key books. The fine authors of these are easily found. *Disasters by Design* offers a compendium about disaster management, as does *Natural Hazard Mitigation*. *Disasters and Democracy* documents history in our dealings with disasters and includes insightful analysis. *Floodplains* is a fine textbook, as is *Floodplain Management*. Practitioner workbooks include *Design for Flooding* and *Greenways: A Guide to Planning, Design, and Development*. A number of official histories published by the Army Corps of Engineers, including *The Army Corps of Engineers and the Evolution of Federal Flood Plain Management Policy*, revealed much. *The Big*

274   SOURCES

Ones is an engaging overview about disasters and our approach to them. *Battling the Inland Sea* tells the flooding history of California's Central Valley, while *Rising Tide* describes the 1927 flood. For narrative on coastal flooding, which I've only inferentially addressed, see *The Water Will Come*. Gilbert White's biography is *Living with Nature's Extremes*. For White's own work, see *Strategies of American Water Management*.

Newspaper and magazine articles formed essential building blocks for this story. Among others, *The New York Times, The Guardian, Los Angeles Times, The Hill, The Sacramento Bee, Houston Chronicle, E&E News,* and *Politico* were especially helpful, as was National Public Radio with broadcasts archived online.

Mary Bart, communications lead at the Association of State Floodplain Managers, produces an excellent newsletter that keeps its readers current on these issues. Key players and interviews there included the organization's co-founder Larry Larson—a floodplain management legend with a lifetime of dedication, knowledge, and wisdom. The Federal Emergency Management Agency's communications staff were helpful in connecting my queries to various specialists, for which the agency requested citation as "FEMA News Desk." I requested full reviews of relevant chapters by FEMA and the Army Corps of Engineers, but unfortunately those requests were unsuccessful, which I regard as an indicator of how overloaded are the schedules of dedicated public officials in those agencies. A list of people with whom I had significant interviews follows. The interviews occurred mostly in 2021–22, though a few date back many years.

- Berginnis, Chad, Association of State Floodplain Managers
- Blackwelder, Brent, formerly of the Environmental Policy Center
- Bouchey, Rebecca, Forterra Foundation
- Bridges, Todd, Army Corps of Engineers, Engineering with Nature program
- Brower Solisti, Kate, niece of Professor Lincoln Brower
- Brown, Chris, former president, American Rivers; also river programs lead, National Park Service and Forest Service
- Cain, John, River Partners
- Canaan, David, director, Mecklenburg County Stormwater Services
- Carey, Bob, The Nature Conservancy, WA
- Coyle, Kevin, former president, American Rivers
- Dorothy, Olivia, American Rivers, Mississippi River program
- Fahlund, Andrew, Water Foundation
- Fowler, David, Association of State Floodplain Managers
- Galloway, Gerald, University of MD; Brigadier General, Army Corps of Engineers

Hoffner, Jenny, American Rivers

Kroening, David, Mecklenburg County Stormwater Services

Larson, Larry, Association of State Floodplain Managers

Lavelle, John, Lycoming County Planning Commission, PA

Leopold, Luna, hydrologist, formerly at University of CA at Berkeley

Lightbody, Laura, Pew Charitable Trusts

Lindsey, Roger, Metro Water Services, Nashville

Love, David, Mecklenburg County Stormwater Services

MacDonald, Clyde, former staff, CA State Legislature

Medberry, Mike, author, Boise

Meral, Jerry, former deputy director, CA Dept. of Water Resources; also executive director, CA Planning and Conservation League

Mierzwa, Michael, state floodplain manager, CA Dept. of Water Resources

Milstein, Michael, National Oceanic and Atmospheric Administration, Seattle

Moore, Rob, Natural Resources Defense Council

Opperman, Jeff, World Wildlife Fund

Palko, Tom, Metro Water Services, Nashville

Parker, Mike, resident, Hamilton, WA

Pinter, Nicholas, University of CA at Davis

Platt, Rutherford, University of MA

Shader, Eileen, floodplain program lead, American Rivers

Shoji, Crystal, professional planner, Coos Bay, OR

Shorin, Bonnie, National Oceanic and Atmospheric Administration, Seattle

Stork, Ron, Friends of the River, CA

Struble, Bob, Brandywine Red Clay Alliance, PA

Swain, Daniel, University of CA at Los Angeles

Thompson, Carol, former floodplain resident, Charlotte, NC

Touton, Camille, former staff, House of Representatives, Water and Power Committee

Trautman, Tim, Mecklenburg County Stormwater Services

Walls, Jerry, former planning director, Lycoming County, PA

Washburn, Tim, Sacramento Area Flood Control Agency

Notes

2. THE ESSENTIAL HISTORY OF HIGH WATER

1. Ingram, B. Lynn and Frances Malamud-Roam. *The West Without Water.* University of California Press, 2013, p 32–34.
2. Kahrl, William L. ed. *The California Water Atlas.* State of California, 1979, p 78.
3. Dettinger, Michael D. and B. Lynn Ingram. "The Coming Megafloods." *Scientific American,* Jan. 2013.
4. Ralph, F. Martin. *Emergence of the Concept of Atmospheric Rivers.* International Atmospheric Rivers Conference, Scripps Institution of Oceanography, Aug. 8, 2016, p 12.
5. Berwyn, Bob. "A Surge from an Atmospheric River Drove California's Latest Climate Extremes." *Inside Climate News,* Feb. 2, 2021, online p 5.
6. Hodges, Glenn. "Did a Mega-Flood Doom Ancient American City of Cahokia?" nationalgeographic.com, Nov. 2, 2013, p 1.
7. Langolis, Krista. "Have We Underestimated the West's Super-Floods?" *High Country News,* Feb. 28, 2017, online p 2.
8. US Geological Survey. "Overview of the ARkStorm Scenario." June 7, 2011, online p 1.
9. Swain, Daniel et al. "Increasing Precipitation Volatility in 21st-Century California." *Nature Climate Change,* Apr. 2018.

10. Tara, Roopinder. "America's Second Biggest Levee System Is Keeping Sacramento Dry—For Now." engineering.com, Feb. 10, 2021.

11. Allen, George H. and Tamlin M. Pavelsky. "Global Extent of Rivers and Streams." *Science,* Aug. 10, 2018, p 585.

12. Brinkley, Douglas. *The Great Deluge.* Harper Collins, 2006, p xix.

13. Ashley, Sharon T. and Walker S. Ashley. "Flood Fatalities in the United States." *Journal of Applied Meteorology and Climatology,* Mar. 2000, p 813.

14. National Oceanic and Atmospheric Administration. *Hurricanes.* NOAA website.

15. Zhong, Raymond. "Fires, Then Floods: Risk of Deadly Climate Combination Rises." *New York Times,* Apr. 1, 2022.

16. Shap, Elena. "Mapping California's 'Zombie' Forests." *New York Times,* Mar. 6, 2023, online p 1.

17. Shank, William H. *Great Floods of Pennsylvania.* Buchart-Horn Engineers, 1972, p 21.

18. McCullough, David. *The Johnstown Flood.* Simon & Schuster, 1968.

19. Platt, Rutherford H. *Disasters and Democracy.* Island Press, 1999, p 3.

20. Moore, Jamie W. and Dorothy P Moore. *The Army Corps of Engineers and the Evolution of Federal Flood Plain Management Policy.* Institute of Behavioral Science, University of Colorado, 1989, p 1.

21. Ellet, Charles. *The Mississippi and Ohio Rivers.* Lippencott, 1853, p 28.

22. Moore, p 4.

23. Arnold, Joseph L. *The Evolution of the 1936 Flood Control Act.* Office of History, US Army Corps of Engineers, 1988, p 3.

24. Barry, John M. *Rising Tide.* Simon & Schuster, 1997, p 317.

25. Natural Resources Committee. *Drainage Basin Problems and Programs,* Government Printing Office, 1937, p 1–5; cited in Welky, David. *The Thousand-Year Flood.* University of Chicago Press, 2011, p 52.

26. Welky, David. *The Thousand-Year Flood.* University of Chicago Press, 2011, p 232–238.

27. Arnold, Joseph L. *The Evolution of the 1936 Flood Control Act.* Office of History, US Army Corps of Engineers, 1988, p v, vii.

28. Palmer, Tim. *Endangered Rivers and the Conservation Movement.* University of California Press, 1986, p 23.

29. Welky, p 67.

30. Maher, Neil M. *Nature's New Deal.* Oxford University Press, 2008, p 164.

31. Platt, p 11.

32. Hundley, Noirris Jr. *The Great Thirst.* University of California Press, 1992, p 161.

33. Blair, David and Timothy Keptner. *Environmental Concerns in Local Floodplain Management.* Department of Environmental Resources, brochure, 1979, p 3.

34. Godschalk, David R. *Natural Hazard Mitigation.* Island Press, 1999, p 92.

35. Larson, Lee W. "The Great USA Flood of 1993." NOAA/National Weather Service, International Association of Hydrological Sciences conference, Anaheim, CA, June 1996.

36. Faber, Scott. "Flood Policy and Management: A Post-Galloway Progress Report." *River Voices,* newsletter of River Network, summer 1997, p 1.

37. Diringer, Elliot. "Floods Warn Valley of Worse to Come." *San Francisco Chronicle,* Mar. 3, 1997.

38. Dolan, Eric Jay. *A Furious Sky.* Norton, 2020, p 308.

39. Bajak, Frank and Lise Olsen. "Hurricane Harvey's Toxic Impact Deeper Than Public Told." Associated Press, Mar. 23, 2018.

40. Crunden, E. A. "North Carolina's Governor Calls for Sweeping, Inclusive Rebuilding Process after Florence." *Think Progress,* Sep. 18, 2018, online p 2.

41. Association of State Dam Safety Officials. *Case Study: St. Francis Dam.* The Association, 2021, online.

42. Gee, Nathaniel. "Case Study: Buffalo Creek Dam." Association of State Dam Safety Officials website, 2022.

43. Landers, Jay. "Michigan Dam Failures Were 'Foreseeable.'" *Civil Engineering Source,* Apr. 11, 2022.

44. Newberger, Emma. "More Dams Will Collapse as Aging Infrastructure Can't Keep Up with Climate Change." cnbc.com, Apr. 21, 2020.

45. Gallagher, Dorothy. "The Collapse of the Great Teton Dam." *New York Times,* Sep. 19, 1976, online p 11.

46. *Teton Dam Failure, Teton River, Idaho, June 5, 1976.* YouTube.

47. Knudson, Tom and Nancy Vogel. "Aging Dams Are Facing New Pressures." *Sacramento Bee,* Nov. 25, 1997, insert, p 6.

48. Langolis, Krista. "Have We Underestimated the West's Super-Floods?" *High Country News,* Feb. 28, 2017, online p 2.

49. Friends of the River et al. *Motion to Intervene with FERC Permitting Process.* Oct. 17, 2005, p 11.

50. Rogers, Paul. "Oroville Dam: Feds and State Officials Ignored Warnings 12 Years Ago." *San Jose Mercury News,* Feb. 12, 2017.

51. Rogers, J. David. *Design Evolution of Oroville Dam 1956-68.* Missouri University of Science and Technology, 2017.

52. Kovach, Tim. "Today's Disasters Are Really Just Yesterday's Unaddressed Vulnerabilities." *Grist,* May 29, 2015.

53. Zhong, Raymond. "The Coming California Megastorm." *New York Times,* Aug. 12, 2022, online p 18.

54. Leslie, Jacques. "In an Era of Extreme Weather, Concerns Grow Over Dam Safety." *Yale Environment 360,* July 9, 2019, online p 1-4.

55. Jacobs, Jeremy P. "Climate Change Erodes Thin Safety Margins at California Dam." *E & E News Greenwire,* May 8, 2017, online p 2.

56. Palmer, Tim. *Endangered Rivers and the Conservation Movement.* University of California Press, 1986, p 231.

57. Wesselman, Eric and Ron Stork. "To Avoid Catastrophe, Don't Build More Dams." *San Francisco Chronicle,* op-ed, Feb. 15, 2017, and interviews.

58. Stork, Ron. Presentation at the South Yuba River Wild and Scenic Film Festival, Jan. 2020.

59. Zhong, Raymond. "The Coming California Megastorm." *New York Times,* Aug. 12, 2022, online p21.

60. McCann, Martin. *Dam Failures in the United States.* National Performance of Dams Program, Stanford University, Sep. 2018, online.

61. FEMA. *Managing Floodplain Development Through the National Flood Insurance Program.* 1998, Chapter 1, p 15.

62. Leslie, Jacques. "In an Era of Extreme Weather, Concerns Grow Over Dam Safety." *Yale Environment 360,* July 9, 2019, online p 2.

63. American Society of Civil Engineers. *Report Card.* The Society, Apr. 2021, online.

64. Holden, Emily. "Thousands of Run-Down US Dams Would Kill People if They Failed, Study Finds." *The Guardian,* May 23, 2020, online p 1.

65. Leslie, Jacque. "In an Era of Extreme Weather, Concerns Grow Over Dam Safety." *Yale Environment 360,* July 9, 2019, online.

66. Mount, Jeffrey. "Floods in California." Pacific Policy Institute website, Nov. 2021.

67. Lieb, David A. "Aging Dams Could Soon Benefit from $7B Federal Loan Program." Associated Press, June 10, 2022.

68. Floodplain Management Association. *2019 California Floodplain Risk Management Symposium.* 2019, p 5.

69. Grossman, Elizabeth. *Watershed: The Undamming of America.* Counterpoint, 2002.

70. Cosier, Susan. "US Farmers Count Cost of Catastrophic 'Bomb Cyclone' in Midwest." *The Guardian,* April 27, 2019, online p 1.

71. Gustin, Georgina. "As Extreme Weather Batters America's Farm Country, Costing Billions, Banks Ignore the Financial Risks of Climate Change." *Inside Climate News,* May 2, 2021, online p 8.

72. Salter, Jim. "Flooding Poses Potential Risk for 1 Million Private Wells." apnews.com, Mar. 25, 2019, p 1.

73. Korda, Matt. "The U.S. Nuclear Deterrent Is Not Prepared for Climate Catastrophe." forbes.com, March 16, 2020, p 1.

74. McGreal, Chris. "'So Much Land Under Water': Extreme Flooding Is Drowning Parts of the Midwest." *The Guardian,* June 2, 2019, p 9.

75. Smith, Mitch and John Schwartz. "'Breaches Everywhere': Flooding Bursts Midwest Levees, and Tough Questions Follow." *New York Times,* Mar. 31, 2019, online p 1.

76. Morford, Stacy. "Rising Temperatures Are Increasing the Hurricane Risks to New Orleans and Other Coastal Area." *Scientific American,* Aug. 30, 2021, online p 3, 5.

77. Sengupdta, Somini. "Why the Wilder Storms? It's a 'Loaded Dice' Problem." *New York Times,* Oct. 5, 2018.

78. Newman, Andy. "As Flooding Death Toll Rises, the Governors of New York and New Jersey Say They Expect Help from Washington." *New York Times,* Sep. 3, 2021.

79. National Oceanic and Atmospheric Administration. "Flood Related Hazards." National Weather Service, June 2021, online.

80. Senator Roger Wicker, in testimony for Senate Bill 558 in 2021. *News & Views.* Association of State Floodplain Managers, Apr. 2021.

81. Staletovich, Jenny. "Shutting an Agency Managing Sprawl Might Have Put More People in Hurricane Ian's Way." National Public Radio, Oct. 8, 2022, online p 5.

82. White, Martha C. "Tens of Thousands Likely Jobless After Hurricane Ian, Economists Say." *Markets Now,* CNN Business online, Oct. 12, 2022, online p 1.

83. Sacks, Brianna. "For Some, Life After Ian Is 'More Tragic Than the Hurricane Itself.'" *Washington Post,* Feb. 1, 2023, online.

84. Vileisis, Ann. *Discovering the Unknown Landscape: A History of America's Wetlands.* Island Press, 1997, p 350.

3. RIVERS NEED FLOODS AND NATURE NEEDS FLOODPLAINS

1. Brower, Lincoln P. *The Flooding River: A Study in Riverine Ecology.* John Wiley and Sons, 16 mm film.

2. Brower, Lincoln. "Water and the Flooding River." *A Sweet Briar College Learning Resource,* 1975, online and interview.

3. McShane, John H. "Shifting the Paradigm for the 21st Century: Protecting and Restoring the Natural Resources and Functions of Floodplains." *Water Resources Impact,* Mar. 2014, p 13.

4. Jakubinsky, Iiri et al. "Managing Floodplains Using Nature-Based Solutions to Support Multiple Ecosystem Functions and Services." *Wires Water: Wiley Interdisciplinary Reviews.* July 3, 2021. (No page number; see heading #3, The Ecosystem Functions and Services of a Floodplain.)

5. Watson, Donald and Michele Adams. *Design for Flooding.* John Wiley & Son, 2011, p 95.

6. Wohl, E. E. "Geomorphic Effects of Floods." In *Inland Flood Hazards: Human, Riparian and Aquatic Communities.* Cambridge University Press, 2000, p 181–212.

7. Freitag, Bob et al. *Floodplain Management.* Island Press, 2009, p 44.

8. Owen, H. James and Glenn R. Wall. *Floodplain Management Handbook.* US Water Resources Council, 1981, p 1. Also FEMA. *A Unified National Program for Floodplain Management.* 1994, p 3.

9. Alex Demas, USGS, personal communication, 2021.

10. Resh, V. H. et al. "The Role of Disturbance in Stream Ecology." *Journal of the North American Benthological Society,* 1988, p 433–455.

11. Opperman, Jeffrey J. *Floodplains.* University of California Press, 2019, p 45, 99.

12. California Department of Water Resources. *Going with the Flow: How Aquifer Recharge Reduces Flood Risk,* Aug. 3, 2022, online p 1.

13. Taylor, Charles A. and Hannah Drukenmiller. "Wetlands, Flooding, and the Clean Water Act." *American Economic Review,* Apr. 4, 2022.

14. Gordon, Brad A. et al. "Nutrient Retention in Ecologically Functional Floodplains: A Review." *Water Review,* Oct. 4, 2020, p 1.

15. Opperman, p 41.

16. Reeves, Gordon H. et al. "Fish and Aquatic Ecosystems of the Oregon Coast Range." In *Forest and Stream Management in the Oregon Coast Range.* Oregon State University Press, 2002, p 79.

17. Norse, Elliott A. *Ancient Forests of the Pacific Northwest.* Island Press, 1990, p 52–53, 105–107.

18. Brinson, M. M. *Riparian Ecosystems: Their Ecology and Status.* USDI Fish and Wildlife Service, 1981, p 155.

19. US Department of Agriculture, Forest Service, Southwestern Region. *Where Land Meets Water,* brochure, no date, p 1.

20. Roman-Palacios et al. "Freshwater Habitats Are Fragile Pockets of Exceptional Biodiversity." *Ecology Letters,* Apr. 22, 2022.

21. Garone, Phil. *The Fall and Rise of the Wetlands of California's Great Central Valley.* University of California Press, 2011, p 31.

22. Goldfarb, Ben. *Eager.* Chelsea Green, 2018, p 35.

23. Cheney, Ed. *Managing Change: Livestock Grazing on Western Riparian Areas.* Environmental Protection Agency, booklet, 1993, p 154–155.

24. Friends of the River. *Beyond Flood Control: Flood Management and River Restoration,* newsletter, 1997, p 4.

25. Robbins, Jim. "Why the World's Rivers Are Losing Sediment and Why It Matters." *Yale Environment 360,* June 20, 2017, p 2.

26. Palmer, Tim. *Field Guide to California Rivers.* University of California Press, 2012, p 235.

27. Personal communication with the Corps' public information officer, Feb. 21, 2022.

28. Williams, Philip & Associates. *An Evaluation of Flood Management Benefits Through Floodplain Restoration on the Willamette River, Oregon.* River Network, 1996, p 3.

29. Parsons, Melissa. "Extreme Floods and River Values: A Social-Ecological Perspective." *River Research Applications,* Wiley Online Library, Sep. 17, 2018.

30. Association of State Floodplain Managers. *Natural and Beneficial Floodplain Functions.* Sep. 16, 2008, p 5.

4. DAMS FOR FLOOD CONTROL: THE PROMISE AND THE REALITY

1. Morgan, Arthur. *Dams and Other Disasters.* Porter Sargent, 1971, p 311.

2. Kelly, Robert. *Battling the Inland Sea.* University of California Press, 1989, p 192.

3. Leopold, Luna and Thomas Maddock Jr. *The Flood Control Controversy.* Ronald Press, 1954, p 142.

4. Moore, Jamie W. and Dorothy P Moore. *The Army Corps of Engineers and the Evolution of Federal Flood Plain Management Policy.* Institute of Behavioral Science, University of Colorado, 1989, p 2.

5. Morgan, p 300.

6. Palmer, Tim. *Youghiogheny: Appalachian River.* University of Pittsburgh Press, 1984, p 81.

7. Powledge, Fred. "Can TVA Change Its Spots?" *Audubon,* Mar. 1983, p 61.

8. Palmer, Tim. *Endangered Rivers and the Conservation Movement.* University of California Press, 1986, p 181.

9. Holmes, Beatrice Hort. *History of Federal Water Resources Programs and Policies, 1961-70.* US Department of Agriculture, 1979, p 76.

10. Palmer, 1986, p 43.

11. Palmer, 1984, p 91.

12. Grinde, Donald A. Jr. "Iroquois Political Theory and the Roots of American Democracy." In *Exiled in the Land of the Free.* Clear Light, 1992, p 277.

13. Josephy, Alvin M. "Cornplanter, Can You Swim." *American Heritage,* Dec. 1968, p 4-9.

14. Morgan, p 339-367.

15. Haveman, Robert H. *Water Resources Investment and the Public Interest.* Vanderbilt University Press, 1965.

16. Arnold, Joseph L. *The Evolution of the 1936 Flood Control Act.* Office of History, US Army Corps of Engineers, 1988, p 91.

17. Platt, Rutherford K. *Land Use and Society.* Island Press, 1996, p 421.

18. White, Gilbert. *Strategies of American Water Management.* Ann Arbor Paperbacks, 1969, p 50.

19. Palmer, 1986, p 185.

20. ACE public affairs office, personal communication, Feb. 17, 2022.

21. Moore, p 105.
22. Rosen, Howard and Martin Reuss eds. *The Flood Control Challenge: Past, Present, and Future.* Public Works Historical Society, 1988, p 104.
23. Palmer, 1986, p 62.
24. Schick, Tony and Irena Hwang. "The US Has Spent More Than $2B on a Plan to Save Salmon. The Fish Are Vanishing Anyway." *In The News,* Oregon Public Broadcasting, Aug. 13, 2022, online.
25. Knudson, Tom and Nancy Vogel. "Aging Dams Are Facing New Pressures." *Sacramento Bee,* insert, Nov. 23–27, 1997, p 3, 6.
26. Hecht, Jeff. "The Incredible Shrinking Mississippi Delta." *New Scientist,* Apr. 1990, p 44.
27. Phillips, Guy. *The New Melones Project: A Review of Current Economic and Environmental Issues.* California Resources Agency, Sep. 1979.
28. Palmer, Tim. *Rivers of California.* Heyday, 2010, p 158.
29. Palmer, 1986, p 128.
30. Interview, 2021.
31. Diringer, Elliot and Ramon G. McLeod. "Floods Warn Valley of Worse to Come." *San Francisco Chronicle,* Mar. 3, 1997, p 8.
32. Knudson, p 3.
33. Knudson, p 2.
34. Williams, Philip. "Inviting Trouble Downstream." *Civil Engineering,* Feb. 1998, p 51. Also Knudson, p 6.
35. Williams, p 52.
36. Haefeli, Laura. "Parts of Sacramento Almost Had Catastrophic Flooding After Pumping Plants Lost Power." *CBS Sacramento,* Jan. 10, 2023, online p 1.
37. Kelly, Tyler J. "The Fight to Tame a Swelling River with Dams That May Be Outmatched by Climate Change." *New York Times,* Mar. 21, 2019, online.
38. Leopold, Luna and Thomas Maddock. *The Flood Control Controversy.* The Conservation Foundation, 1954, p ix.
39. Faber, Scott. "The Great Flood of 1993." *River Voices,* newsletter of River Network, winter 1994, p 5.
40. Williams, Phil. "Here Comes the Flood," *World Rivers Review,* fourth quarter 1993, p 8.
41. Owen, David. "The Biggest Potential Water Disaster in the United States." *New York Times,* May 11, 2022.
42. Stanway, David. "World's Dams to Lose a Quarter of Storage Capacity by 2050." Reuters, Jan. 11, 2023, online p 1.
43. Palmer, 1986, p 230.
44. Knudson, p 8.
45. Knudson, p 11.

5. BROKEN BARRIERS

1. Rosen, Howard and Martin Reuss eds. *The Flood Control Challenge: Past, Present, and Future.* Public Works Historical Society, 1988, p 89.
2. Hinshaw, Robert. *Living with Nature's Extremes.* Johnson Books, 2006, p xiii.
3. Kelley, Tyler J. "The Fight to Tame a Swelling River with Dams That May Be Outmatched by Climate." *New York Times,* Mar. 21, 2019, online.
4. American Rivers. *In Harm's Way: The Costs of Floodplain Development.* American Rivers, brochure, 1999, p 11.
5. Army Corps of Engineers. "USACE Opens Updated National Levee Database." June 5, 2018, online. Also personal communication with the Corps public affairs office, Feb. 21, 2022.
6. National Committee on Levee Safety. *Draft Recommendations for a National Levee Safety Program: A Report to Congress,* 2009.
7. Personal communication with the Corps public affairs office, Feb. 21, 2022.
8. American Society of Civil Engineers. *Report Card.* The Society, Apr. 2021, online.
9. Floodplain Management Association. *2019 California Floodplain Risk Management Symposium.* White paper, 2019, p 3.
10. Friends of the River. *Beyond Flood Control: Flood Management and Restoration.* Friends of the River, booklet, 1997, p 7.
11. Sabalow, Ryan. "Levee 'Armoring' Along the American River Parkway Draws Concerns." *Sacramento Bee,* Mar. 25, 2016.
12. Ingram, Lynn and Frances Malamud-Roam. *The West Without Water.* University of California Press, 2013, p 209.
13. Interview with Tim Washburn, 2021.
14. California Department of Water Resources. *Flood Control Systems Status Report.* Central Valley Flood Management Program, 2011.
15. Dettinger, Michael D. and B. Lynn Ingram. "The Coming Megafloods." *Scientific American,* Jan. 2013.
16. Song, Lisa and Patrick Michaels. "Flood Thy Neighbor: Who Stays Dry and Who Decides?" *ProPublica,* Feb. 10, 2021, online p 3.
17. Steinberg, Ted. *Acts of God.* Oxford University Press, 2000, p xvii.
18. Kelley, Robert. *Battling the Inland Sea.* University of California Press, 1989, p x.
19. Smith, Mitch. "Midwestern Floods Pit Communities Against One Another as Levees Rise Ever Higher." *New York Times,* May 7, 2019, online p 1.
20. Forbs, J. B. "Levee Wars: Missouri Residents Complain Illinois Levee Was Built Higher Without Approval." *St. Louis Post-Dispatch,* Sep. 21, 2015, online p 7, 10.

286   NOTES TO PAGES 104–111

21. James, Ian and Suzanne Rust. "Worry and Suspicion Reign as Once-Dry Tulare Lake Drowns California Farmland." *Los Angeles Times,* Mar. 23, 2023, online.

22. Belt, Charles Jr. "The 1973 Flood and Man's Constriction of the Mississippi River." *Science,* Aug. 1975 reprint, p 1.

23. Faber, Scott. *On Borrowed Land.* Lincoln Institute, 1996, p 6.

24. "Levee Plan Is Attacked." *St. Charles Journal,* Oct. 4, 1971, quoted in Steinberg, Ted. *Acts of God.* Oxford University Press, 2000, p 102.

25. Cusick, Daniel. "Record Floods Worsened by Warming and Levees. 'How idiotic.'" *E & E News.* May 7, 2019, online p 3, 4.

26. Heine, Reuben A. and Nicholas Pinter. "Levee Effects upon Flood Levels: An Empirical Assessment." *Hydrological Processes,* Oct. 2012.

27. Neeley, Todd. "Conservation Groups Allege Mississippi River Floods Caused by River-Training Structures." *Progressive Farmer,* Apr. 13, 2020, online p 1.

28. Vileisis, Ann. *Discovering the Unknown Landscape: A History of America's Wetlands.* Island Press, 1997, p 245–251.

29. Palmer, Tim. *Lifelines.* Island Press, 2013, p 157.

30. Doyle, Martin. *The Source.* W. W. Norton, 2018, p 272.

31. Martin, Glen. "Trouble When Rain Comes: California's Concrete Flood Channels Are Decaying." *California Magazine,* Dec. 6, 2013, online p 3.

32. Zhong, Raymond. "The Coming California Megastorm." *New York Times,* Aug. 12, 2022, p 34.

33. Platt, Rutherford H. *Land Use and Society.* Island Press, 1996, p 422.

34. International Society of Soil Mechanics and Geotechnical Engineering, Technical Committee on Geotechnical Aspects of Dikes and Levees. *Failure Paths for Levees.* Feb. 2022, online.

35. Polly, Kris. "Developing Flood Protection in California's Capital." *Municipal Water Leader,* June 15, 2018, online p 2.

36. Association of State Floodplain Managers Foundation. *2019 California Floodplain Risk Management Symposium.* The Association, white paper, 2019, p 10.

37. Kelley, p 205.

38. Diringer, Elliot. "Latticework of Levees Stretched to the Breaking Point." *San Francisco Chronicle,* Mar. 3, 1997, p A9.

39. Opperman, Jeffrey J. *Floodplains.* University of California Press, 2019, p 119.

40. Kelley, p 310.

41. Green, Dorothy. *"Managing Water: Avoiding Crisis in California."* University of California Press, 2007, p 49, 50. Also Delta Stewardship Council. *Delta Adapts: Creating a Climate Resilient Future.* 2021, online.

42. Faber, p 6.

43. Freitag, Bob et al. *Floodplain Management.* Island Press, 2009, p 14.
44. Smith, p 2, 4.
45. Smith, Mitch and John Schwartz. "'Breaches Everywhere': Flooding Bursts Midwest Levees, and Tough Questions Follow." *New York Times,* Mar. 31, 2019, online p 1.
46. Twain, Mark. *Life on the Mississippi.* 1883, Chapter 28.
47. Mount, Jeffrey F. *California Rivers and Streams.* University of California Press, 1995, p 306.
48. National Research Council. *Levee Policy for the National Flood Insurance Program.* National Academy Press, 1982, Introduction, p 2.
49. American Society of Civil Engineers. "Overview of Levees." In *Report Card for America's Infrastructure,* 2021.
50. Smith, p 2.
51. Schleifstein, Mark. "Almost 2/3 of Americans Live in Counties with Levees, FEMA Data Shows." *The Times-Picayune,* nola.com, Aug. 11, 2020, p 4, 5.
52. Smith, p 2.
53. Flavelle, Christopher. "With More Storms and Rising Seas, Which U.S. Cities Should Be Saved First?" *New York Times,* June 10, 2019, online.
54. Dolin, Eric Jay. *A Furious Sky.* Liveright, 2020, p 275.
55. Personal communication with ACE public affairs office, Feb. 21, 2022.
56. Haefeli, Laura. "Parts of Sacramento Almost Had Catastrophic Flooding After Pumping Plants Lost Power." *CBS Sacramento,* Jan. 10, 2023, online p 1.
57. Diringer, Elliot and Ramon G. McLeod. "Floods Warn Valley of Worse to Come." *San Francisco Chronicle,* Mar. 3, 1997.
58. Mileti, Dennis S. *Disasters by Design.* Joseph Henry Press, 1999, p 25.
59. Becker, William S. *The Creeks Will Rise.* Chicago Review Press, 2021, p 128.
60. Lohan, Tara. "Let Rivers Flood: Communities Adopt New Strategies for Resilience." *The Revelator,* July 8, 2019, online p 2, 3.
61. Sommer, Lauren. "Where Levees Fail in California, Nature Can Step In to Nurture Rivers." National Public Radio, Mar. 29, 2017, online p 3, 4.
62. Woolington, Josephine. "Set It Back: Moving Levees to Benefit Rivers, Wildlife and Communities." *The Revelator,* June 21, 2022.
63. Christensen, Jon. "A Flood of Lessons." *Nature Conservancy,* July/Aug. 1997, p 9.
64. Interview, 2021.
65. Zhong, Raymond. "The Coming California Megastorm." *New York Times,* Aug. 12, 2022, online p 33.
66. Knudson, Tom and Nancy Vogel. "Momentum Slowly Builds for Reforms." *The Sacramento Bee,* insert, Nov. 23–27, 1997, p 11.

6. HIGHER FLOODS AND THE ENDLESS STORM

1. Ingram, B. Lynn and Frances Malamud-Roam. *The West Without Water.* University of California Press, 2013, p 147.
2. Gleick, Peter H. "Flooding in California Has Been Worse Before—and Likely Will Be Again." *New York Times,* Jan. 10, 2023, online op-ed.
3. Jones, Lucy. *The Big Ones.* Anchor Books, 2018, p 75.
4. Kaplan, Sarah. "'We Need Help': North Carolina Towns Plead for Dam, Levee Upgrades." *The Washington Post,* Sep. 19, 2018.
5. Sommer, Lauren. "An Unexpected Item Is Blocking Cities' Climate Change Prep: Obsolete Rainfall Records." National Public Radio, Feb. 9, 2022.
6. Oakford, Samuel et al. "America Underwater." *Washington Post,* Dec. 12, 2022, online p 5.
7. Mott, Nick. "Rising Rivers Don't Necessarily Follow the Lines on a Map" *High Country News,* Sep. 12, 2022, online p 3, 6.
8. Hersher, Rebecca. "Living in Harm's Way: Why Most Flood Risk Is Not Disclosed." National Public Radio, Oct. 20, 2020, online p 4.
9. Flavelle, Christopher et al. "New Data Reveals Hidden Flood Risk Across America." *New York Times,* June 29, 2020.
10. FEMA. *Managing Floodplain Development Through the National Flood Insurance Program.* 1998, Chapter 1, p 22.
11. Freitag, Bob et al. *Floodplain Management.* Island Press, 2009, p 45.
12. Luoma, Jon R. *The Hidden Forest.* Henry Holt & Co., 1990, p182.
13. Bruggers, James. 'Strip Mining Worsened the Severity of Deadly Kentucky Floods." *Inside Climate News,* Aug. 7, 2011, online p 4.
14. Smith, Hayley. "As Drought Lingers, Larger and More Destructive Wildfires Pose New Threats to Water Supply." *Los Angeles Times,* Feb. 21, 2022.
15. Touma, Danielle et al. "Climate Change Increases Risk of Extreme Rainfall Following Wildfire in the Western United States." *Science Advances,* Apr. 1, 2002, p 1.
16. "CNN State of the Union," Sep. 4, 2022, cited by Society of Environmental Journalists, Sep. 7, 2022, online.
17. McKibben, Bill. *Falter: Has the Human Game Begun to Play Itself Out?* Henry Holt, 2019, p 24.
18. Peterson, T. C. et al. *U.S. Climate Extremes Index.* 2013.
19. McKenna, Phil. "Extreme Weather Flooding the Midwest Looks a Lot Like Climate Change." *Inside Climate News,* May 8, 2017.
20. Milman, Oliver. "Disasters Like Louisiana Floods Will Worsen as Planet Warms, Scientists Warn." *The Guardian,* Sep. 1, 2016, online p 1.
21. Basu, Brishti et al. "This Map Shows Which U.S. Homes Will Flood Over the Next 30 Years Due to Climate Change." vice.com, June 29, 2020, online p 1.

22. Knudson, Tom and Nancy Vogel, "The Gathering Storm." *Sacramento Bee*, Nov. 23, 1997, p 2.

23. Joyce, Christopher. "Scientists Glimpse Houston's Flooded Future in Updated Rainfall Data." National Public Radio, Nov. 3, 2017.

24. Interview, Sep. 29, 2021.

25. Knudson, Tom. "Flooding: Push to Build in Harm's Way." *Sacramento Bee*, insert, Nov. 23–27, 1997, p 2.

26. D'Angelo, Chris. "Climate Change Has Loaded the Dice on the Frequency of 100 Year Floods." *Huffington Post*, Aug. 30, 2017.

27. Joyce, Christopher. "Scientists in Houston Tell a Story of Concrete, Rain and Destruction." National Public Radio, Morning Edition, Nov. 9, 2017, online p 5.

28. Lakhani, Nina. "US Hit by 20 Separate Billion-Dollar Climate Disasters in 2021, NOAA Report Says." *The Guardian*, Jan. 11, 2022.

29. McDaniel, Eric. "Carbon Dioxide, Which Drives Climate Change, Reaches Highest Level in 4 Million Years." National Public Radio, June 7, 2021, online.

30. Fountain, Henry and Mira Rojanasakul. "The Last 8 Years Were the Hottest on Record." *New York Times*, Jan. 10, 2023, online p 2.

31. YCC Team. "Pacific Northwest Heat Wave 'Virtually Impossible' Without Global Warming, Scientists Find." *Yale Climate Connections*, Nov. 2, 2021.

32. NOAA. "It's Official: July Was Earth's Hottest Month on Record." NOAA.gov news, Aug. 13, 2021, online p 1.

33. "Risk of Extreme Weather from Climate Change to Rise Over Next Century, Report Says." *The Lancet*, June 21, 2015.

34. Jamail, Dahr. *The End of Ice*. The New Press, 2019, p 116.

35. Readfearn, Graham. "Amount of Ocean Heat Found to Be Accelerating and Fueling Extreme Weather Events." *The Guardian*, Oct. 18, 2022, online p 1.

36. Harvey, Fiona. "Record Ocean Temperatures Put Earth in 'Uncharted Territory', Say Scientists." *The Guardian*, Apr. 26, 2023, online p 1.

37. US Global Change Research Program. "Sea Level Rise." GlobalChange.gov, 2019, online p 1.

38. NOAA. *2022 Sea Level Rise Technical Report*. Feb. 2022.

39. Borenstein, Seth. "US Could See a Century's Worth of Sea Rise in Just 30 Years." Associated Press, Feb. 15, 2022.

40. Colgan, William T. et al. "Greenland Ice Sheet Climate Disequilibrium and Committed Sea-Level Rise." *Nature Climate Change*, Aug. 29, 2022, p 1.

41. Baloch, Shah Meer. "Pakistan Floods Death Toll Passes 1,000, Say Officials." *The Guardian*, Aug. 28, 2022, online p 1.

42. Ripple, William J. et al. "World Scientists' Warning of a Climate Emergency." *BioScience*, Jan. 2020, p 8–12.

43. Rojanasakul, Mira and Catrin Einhorn. "Which Elected Leaders Should Do More on Climate? Here's What Americans Say." *New York Times*, July 19, 2022, online p 1.

44. Jones, Nicola. "Wild Weather and Climate Change," *Yale Environment 360*, May 9, 2017.

45. Waldman, Scott. "Global Warming Tied to Hurricane Harvey." *Scientific American*, Dec. 14, 2017, online.

46. Somer, Lauren. "An Unexpected Item Is Blocking Cities' Climate Change Prep: Obsolete Rainfall Records." National Public Radio, Feb. 9, 2022, p 5.

47. US Global Change Research Program. *Fourth National Climate Assessment, Vol. 2.* 2018, fig. 2.6.

48. Prein, Andreas F. et al. "The Future Intensification of Hourly Precipitation Extremes." *Nature Climate Change*, Dec. 5, 2016, p 1.

49. Bane, Brendan. "Wettest Winter Storms in the Western U.S. Growing Wetter." Pacific Northwest National Laboratory News and Media, online p 2.

50. Harp, Ryan et al. "Observed Changes in Daily Precipitation Intensity in the United States." *Geophysical Research Letters*, Sep. 27, 2022, p 1.

51. IPCC. "Summary for Policymakers." In *Climate Change 2021: The Physical Science Basis.* 2021, B2.2, p 19.

52. Berwyn, Bob. "A Surge from an Atmospheric River Drove California's Latest Climate Extremes." insideclimatenews.org, Feb. 1, 2021, online p 4.

53. Sahagun, Louis. "Risk of Catastrophic California 'Megaflood' Has Doubled Due to Global Warming, Researchers Say." *Los Angeles Times*, Aug. 12, 2022, p 2, 3. Also, Swain, Daniel and Xingying Huang. "Climate Change Is Increasing the Risk of a California Megaflood." *Science Advances*, Aug. 12, 2022, p 1.

54. Dolin, Eric Jay. *A Furious Sky*, using NOAA data. Liveright, 2020, p 308.

55. Thompson, Stuart A. and Yaryna Serkez. "Every Place Has Its Own Climate Risk. What Is It Where You Live?" *New York Times*, Sep. 28, 2020, online p 2.

56. Mattise, Jonathan and Jeffrey Collins. "22 Dead, Many Missing after 17 Inches of Rain in Tennessee." Associated Press, Aug. 22, 2021, online p 2.

57. Ingram, p 152.

58. Gustin, Georgina. "As Extreme Weather Batters America's Farm Country, Costing Billions, Banks Ignore the Financial Risks of Climate Change." *Inside Climate* News, May 2, 2021, online p 2.

59. Lustgarten, Abraham. "Climate Change Will Force a New American Migration." *ProPublica*, Sep. 15, 2020, p 13.

60. Toohey, Grace. "No One Expected 31 Atmosphereic Rivers Storms to Hit California." *Los Angeles Times*, Apr. 11, 2023, online p 1.

61. Sweet, William V. et al. "Patterns and Projections of High Tide Flooding Along the U.S. Coastline Using a Common Impact Threshold," NOAA Technical Report, Feb. 2018.

62. Kossin, James P. "A Global Slowdown of Tropical-Cyclone Translation Speed." *Nature*, June 6, 2018, p 104.

63. Ingram, p 193.
64. Berwyn, Bob. "Destructive Flood Risk in U.S. West Could Triple if Climate Change Left Unchecked." *Inside Climate News,* Aug. 6, 2018, online p1.
65. Northey, Hannah. "The Flood Insurance Debate Returns. Here's What to Expect." *EE News,* Apr. 27, 2021, online p 4.
66. Wing, Oliver E. J. et al. "Inequitable Pattern of US Flood Risk in the Anthropocene." *Nature Climate Change,* Feb. 2022, p 157.
67. Swain, D. L. et al. "Increased Flood Exposure Due to Climate Change and Population Growth in the United States." *Earth's Future,* Oct. 30, 2020, p. 1, 14.
68. Nerem, R. S. et al. "Climate-Change-Driven Accelerated Sea-Level Rise." *Proceedings of the National Academy of Sciences,* Feb. 27, 2018, online p 1.
69. FEMA. *The Impact of Climate Change and Population Growth on the National Flood Insurance Program through 2100.* 2013.
70. First Street Foundation. *Nationwide Resilience Report Finding 25% of All Critical Infrastructure and 23% of Roads Have Flood Risk.* Press release, Oct. 11, 2021.
71. Shao, Elena. "How Is Climate Affecting Floods?" *New York Times,* July 26, 2022, online p 1.
72. Interview, Dr. Daniel Swain, July 27, 2022.
73. Weaver, Scott. "We Just Had Five 1,000-Year Floods in Less Than a Year." Environmental Defense Fund website, Sep. 1, 2016. Also Cappucci, Matthew. "Five 1,000-Year Flood Events in Five Weeks." *Washington Post,* Aug. 23, 2022.
74. Davenport, Coral. "Key to Biden's Climate Agenda Likely to Be Cut Because of Manchin Opposition." *New York Times,* Oct. 15, 2021. Also Goodell, Jeff. "Manchin's Coal Corruption Is So Much Worse Than You Knew." *Rolling Stone,* Jan. 10, 2022.
75. McGrath, Matt. "IPCC: Climate Change Report to Sound Warning on Impacts." *British Broadcasting Corporation News,* Feb. 21, 2022.
76. Holden, Emily. "How the Oil Industry Has Spent Billions to Control the Climate Change Conversation." *The Guardian,* Jan. 8, 2020, online p 1.
77. Jarvis, Brooke. "The Deepest Dig." *The California Sunday Magazine,* Nov. 2, 2014, online.

7. FLOODPLAINS ARE FOR FLOODS

1. US Soil Conservation Service. *1982 National Resource Inventory.* Statistical Bulletin No. 756, 1987, Chapter 1 p 3. Also Blair, David. *Environmental Concerns in Local Floodplain Management.* Pennsylvania Department of Environmental Resources, 1979, p 9.
2. Columbia Climate School. "Making Room for Rivers: A Different Approach to Flood Control." State of the Planet website, June 7, 2011, p 1.

3. NOAA. "Billion-Dollar Weather and Climate Disasters." NOAA website, July 11, 2022, p 1.

4. Interview, 2022.

5. Root, Tik. "Tens of Millions of People Have Been Moving into Flood Zones, Satellite Imagery Shows." *Washington Post,* Aug. 4, 2021, online p 1.

6. Hinshaw, Robert E. *Living with Nature's Extremes.* Johnson Books, 2006, p 28.

7. Interview, Larry Larson, Sep. 14, 2021, from his personal conversations with White.

8. White, Gilbert. *Human Adjustment to Floods.* Doctoral dissertation, University of Chicago, 1942, p 2.

9. American Association of Geographers. "Gilbert White, 1912-2006." The Association, obituary, online, no date.

10. Moore, Jamie W. and Dorothy P Moore. *The Army Corps of Engineers and the Evolution of Federal Flood Plain Management Policy.* Institute of Behavioral Science, University of Colorado, 1989, p 37, 45, 66.

11. White, Gilbert F. *Strategies of American Water Management.* University of Michigan Press, 1969, p 47.

12. Palmer, Tim. *Endangered Rivers and the Conservation Movement.* University of California Press, 1986, p 165.

13. Hinshaw, 139.

14. Moore, p 43, 45.

15. Moore, p 106-111.

16. Rosen, Howard and Martin Reuss eds. *The Flood Control Challenge: Past, Present, and Future.* Public Works Historical Society, 1988, p 97.

17. Holmes, Beatrice Hort. *History of Federal Water Resources Programs and Policies, 1961-70.* US Department of Agriculture, 1979, p 130.

18. McHarg, Ian. *Design with Nature.* Natural History Press, 1969, p 20.

19. Knudson, Tom and Nancy Vogel. "Bad Land-Use Policies Invite Disaster." *Sacramento Bee,* insert Nov. 23-27, 1997, p 5.

20. Sullivan, Kate and Maegan Vazquez. "Biden Surveys Flood Damage in Kentucky and Pledges Federal Support." CNN, Aug. 8, 2022, online p 3.c.

21. Editors. "Awash in Tax Dollars." *News & Observer,* Raleigh, NC, Nov. 11, 1997, p 14A.

22. McShane, John. "Shifting the Paradigm for the 21st Century: Protecting and Restoring the Natural Resources and Functions of Floodplains." *Water Resources Impact,* Mar. 2014, p 1.

23. Interview, 2022.

24. Moore, p 124. Also interview, Robert Zimmerman, Charles River Watershed Association, 2021.

25. Moore, p 130, 134.

26. Interview, Larry Larson, 2021.

27. Interagency Floodplain Management Review Committee. *Sharing the Challenge: Floodplain Management into the 21st Century.* Interagency Floodplain Management Review Committee, June 1994.

28. Interview, 2022.

29. Godschalk, David R. *Natural Hazard Mitigation.* Island Press, 1999, p 30, 66.

30. Frank, Thomas. "Floods Prompt Scrutiny of Missouri River Dams and Levees." *E & E News,* May 20, 2019, online p 3.

31. Platt, Rutherford H. *Disasters and Democracy.* Island Press, 1999, p 93.

32. Holmes, p 27.

33. Leopold, Luna and Thomas Maddock. *The Flood Control Controversy.* The Conservation Foundation, 1954, p 239.

34. Mileti, Dennis S. *Disasters by Design.* Joseph Henry Press, 1999, p 158.

35. Sachs, Noah and David Flores. *Toxic Floodwaters: The Threat of Climate-Driven Chemical Disaster in the James River Watershed.* The Center for Progressive Reform, 2019, p 1.

36. Coleman, Zack. "The Toxic Waste Threat That Climate Change Is Making Worse." *Politico,* Aug. 26, 2019.

37. Tabuchi, Hiroko et al. "Floods Are Getting Worse, and 2,500 Chemical Sites Lie in the Water's Path." *New York Times,* Feb. 6, 2018.

38. Cabrera, Yvette. "Flooding Could Expose Toxic Soil in City Neighborhoods." *Center for Public Integrity Watchdog Newsletter,* July 22, 2022, online p 3.

39. Oglesby, Cameron. "Hog Farms in North Carolina's 100-Year Flood Plain." *Environmental Health News,* May 17, 2021, online p 1–3.

40. Newsome, Melba. "Chicken Frenzy: A State Awash in Hog Farms Faces a Poultry Boom." *Yale Environment 360,* Feb. 14, 2022, p 3.

41. Platt, Rutherford H. *Disasters and Democracy.* Island Press, 1999, p 146–148, 157, 281.

42. Mileti, Dennis S. *Disasters by Design.* Joseph Henry Press, 1999, p 151.

43. Kusler, Jon A. *Government Liability and No Adverse Impact Floodplain Management.* Association of State Floodplain Managers, May 2004, p 2.

44. Diringer, Elliot and Ramon G. McLeod. "Vulnerable Buildings in Harm's Way." *San Francisco Chronicle,* Mar. 3, 1997, p A9.

45. Revkin, Andrew. "In Texas, the Race to Build in Harm's Way Outpaces Flood-Risk Studies and Warming Impacts." *New York Times,* May 26, 2015, online.

46. Interview, 2022.

47. Interview, 2022.

48. NOAA. *Coastal County Snapshots,* 2014, online.

49. Flavelle, Christopher. "New U.S. Strategy Would Quickly Free Billions in Climate Funds." *New York Times,* Jan. 25, 2021, online p 2.

50. Maciag, Mike. '"Building Homes in Flood Zones: Why Does This Bad Idea Keep Happening?" *Governing,* July 27, 2018.
51. Mileti, p 18.
52. Abrams, Amanda. "Climate-Change Risks Get the Attention of Real Estate Investors." *New York Times,* Apr. 19, 2022, p 2.

8. THE INSURANCE CONNECTION

1. National Weather Service. "Flood Safety Awareness Week." Website, 2022.
2. FEMA News Desk, personal communication, July 1, 2022.
3. Moore, Jamie W. and Dorothy P. Moore. *The Army Corps of Engineers and the Evolution of Federal Flood Plain Management Policy.* Institute of Behavioral Science, University of Colorado, 1989, p 42.
4. US Congress. *A Unified National Program for Managing Flood Losses.* House Document 465, 89th Congress, 2nd Session, p 17–18. In Platt, Rutherford. *Disasters and Democracy.* Island Press, 1999, p 77.
5. Conrad, David. "National Flood Insurance Program." *River Voices,* newsletter of River Network, winter 1994, p 6.
6. Moore, p 74, 75.
7. Welky, David. *The Thousand-Year Flood.* University of Chicago Press, 2011, p 282.
8. Hinshaw, Robert E. *Living with Nature's Extremes.* Johnson Books, 2006, p 174.
9. Platt, Rutherford. *Disasters and Democracy.* Island Press, 1999, p 94, 97.
10. Federal Interagency Task Force on Floodplain Management. *Floodplain Management in the United States: An Assessment Report Vol. 1.* Natural Hazards Information Center, 1992, Appendix F, p 4. In Platt, p 82.
11. Moore, p 113.
12. Platt, p 94–99, 112, 126.
13. Platt, p 39.
14. Freitag, Bob et al. *Floodplain Management.* Island Press, 2009, p 87.
15. US House of Representatives. *Report of the Bipartisan Task Force on Disasters.* Dec. 14, 1994, p 8.
16. Godschalk, David R. *Natural Hazard Mitigation.* Island Press, 1999, p 28.
17. Pew Charitable Trusts. "How States Can Manage the Challenges of Paying for Natural Disasters." Pew website, issue brief, Sep. 16, 2020, online p 3.
18. Flavelle, Christopher. "How Government Decisions Left Tennessee Exposed to Deadly Flooding." *New York Times,* Aug. 2, 2021, online p 2.
19. Henson, Bob. "Why Is It So Hard to Fix the National Flood Insurance Program?" www.wunderground.com, Nov. 9, 2017, p 3.

20. FEMA News Desk, personal communication. Also Tanweer, Hira. "How Much Does Flood Insurance Cost?" *QuoteWizard,* June 7, 2021, online.
21. Frank, Thomas. "Most States Are Failing on Building Codes, FEMA Says." *E & E News,* Apr. 6, 2022.
22. Jones, Lucy. *The Big Ones.* Anchor Books, 2019, p 154.
23. Dolin, Eric Jay. *A Furious Sky.* Norton, 2020, p 253, 271.
24. Freitag, p 146.
25. Barry, John M. "The Pandemic Could Get Much, Much Worse. We Must Act Now." *New York Times,* July 14, 2020, online p 1.
26. First Street Foundation. "Defining America's Part, Present, and Future Flood Risk." *Flood Factor Tool,* 2022, online.
27. Association of State Floodplain Managers. "New Tool Helps Property Owners Understand and Lower Flood Risk." *News & Views,* May 13, 2022.
28. Association of State Floodplain Managers. *Flood Mapping for the Nation.* 2020.
29. Berginnis, Chad. "Berginnis Testifies at NFIP Reauthorization Hearing." *News & Views,* Association of State Floodplain Managers, June 2021, online.
30. Larson, Larry. "Policy Matters!" *News & Views,* Association of State Floodplain Managers newsletter, Apr. 2021, and interview, Sep. 14, 2021.
31. Interview, 2021.
32. Moore, p 102.
33. Kusler, Jon and Larry Larson. "Beyond the Ark: A New Approach to U.S. Floodplain Management." *River Voices,* newsletter of River Network, winter 1994, p 3, 12.
34. Wing, Oliver E. J. et al. "Estimates of Present and Future Flood Risk in the Conterminous United States." *Environmental Research Letters,* Feb. 28, 2018, p 1.
35. Association of State Floodplain Managers. *Flood Mapping for the Nation.* Association of State Floodplain Managers, Jan. 2020, p 7, 17, 18.
36. FEMA News Desk, personal communication, Mar. 11, 2022.
37. Kann, Drew. "Flood Risk Is Growing for US Homeowners Due to Climate Change." CNN, Mar. 16, 2021, online p 3.
38. Wright, James M. "Effects of the Flood on National Policy." In *The Great Flood of 1993,* Changnon, Stanley A. ed. Westview Press, 1996, p 69.
39. Hersher, Rebecca. "Living in Harm's Way: Why Most Flood Risk Is Not Disclosed." National Public Radio, Sep. 20, 2020, online p 3.
40. America Adapts, Climate Change Podcast. "Adapting to Chronic Flooding: Survivors' Stories and Actions They Take!" Climate change podcast, June 6, 2022.
41. Floodplain Management Association. *2019 California Floodplain Risk Management Symposium.* The Association and Foundation, 2019, p 7.
42. Hersher, Rebecca. "Living in Harm's Way: Why Most Flood Risk Is Not Disclosed." *All Things Considered,* National Public Radio, Oct. 20, 2020, online.

43. Bart, Mary. *New Resource on State Flood Risk Disclosure Requirements.* Association of State Floodplain Managers, July 20, 2022.

44. Frank, Thomas. "Flood-Battered N.J. Poised to Enact Model Disclosure Law." *Climatewire,* Apr. 14, 2023, online p 1.

45. FEMA. *Partnership for a Safer Future.* Strategic Plan, 1997. In Godschalk, David R. *Natural Hazard Mitigation.* Island Press, 1999, p 59.

46. Flavelle, Christopher. "Homes Are Being Built the Fastest in Many Flood-Prone Areas, Study Finds." *New York Times,* July 31, 2019, online p 1–5.

47. Kann, Drew. "Flood Risk Is Growing for US Homeowners Due to Climate Change." cnn.com, Mar. 16, 2021, p 2.

48. Hunn, David et al. "Build, Flood, Rebuild: Flood Insurance's Expensive Cycle." *Houston Chronicle,* Dec. 7, 2017, p 1–4.

49. Open Secrets: Following the Money in Politics. "Top Spenders." 2022, online p 1.

50. Greer, R. Dell. "Federal Flood Insurance." *Outdoor Recreation Action,* spring 1976, p 16.

51. Warmbrodt, Zachary and Theodoric Meyer. "How Washington Lobbyists Fought Flood Insurance Reform." *Politico,* Sep. 2, 2017, p 4.

52. Flavelle, Christopher and John Schwartz. "Cities Are Flouting Flood Rules. The Cost: $1 Billion." *New York Times,* Sep. 24, 2021, online p 1.

53. Interview, 2022.

54. FEMA News Desk spokesperson, personal communication, June 30, 2022.

55. Kann, Drew. "Flood Risk Is Growing for US Homeowners Due to Climate Change." cnn.com, Mar. 16, 2021, p 1, 2.

56. Hinshaw, p 181.

57. Kidlow, Judith and Jason Scorse. "End Federal Flood Insurance." *New York Times,* Nov. 28, 2012, online p 1.

58. Collins, Gail and Bret Stephens. "This Is Why Politicians Like to Change the Subject." *New York Times,* May 1, 2023, online.

59. Larson, Larry. "Setting People Straight on Six NFIP Myths." *News & Views,* Association of State Floodplain Managers, Aug. 24, 2017, p 1, 2, and interview, 2021.

60. Interview, 2022.

61. FDR. Nomination acceptance speech, June 26, 1936.

62. Kelly, Mary Louise. "Many People Living in Flood-Prone Areas Can't Afford Expensive Flood Insurance." National Public Radio, June 7, 2019, online p 2.

63. Editors. "Hold Strong on Flood Insurance." *Washington Post,* Feb. 2, 2014, p 1, 2.

64. Hunn, David et al. "Build, Flood, Rebuild: Flood Insurance's Expensive Cycle." *Houston Chronicle,* Dec. 7, 2017, p 9, 10.

65. Gaul, Gilbert M. "How Rising Seas and Coastal Storms Drowned the U.S. Flood Insurance Program." *Yale Environment 360*, May 23, 2017.
66. Walsh, Mary Williams. "A Broke, and Broken, Flood Insurance Program." *New York Times*, Nov. 4, 2017, online p 3.
67. Debonis, Mike. "Congress Passes Flood Insurance Extension, Again Punting on Reforms." *Washington Post*, July 31, 2018, online p 2.
68. Faber, Scott. *On Borrowed Land: Public Policies for Floodplains*. Lincoln Institute, 1996, p 1.
69. Insurance Information Institute. *Facts+Statistics: Flood Insurance*, 2021, online.
70. Interview, 2022.
71. Wright, James. "Effects of the Flood on National Policy." In Changnon, Stanley A. ed. *The Great Flood of 1993*. Westview Press, 1996, p 245.
72. Frazin, Rachel. "FEMA Unveils New Flood Insurance Calculation It Says Will Be More Equitable." *The Hill*, Apr. 1, 2021, online p 1.
73. Frank, Thomas. "Almost No One in Ky. Has Flood Insurance, Hindering Recovery." *E & E News*, Aug. 11, 2022, online p 2.
74. FEMA. "Federal Disaster Assistance to Eastern Kentucky Flood Survivors Tops $159 Million." FEMA website, Mar. 9, 2023.

9. MOVING TO HIGHER GROUND

1. Cusick, Daniel. "The U.S. Needs to Address Its Climate Migration Problem." *E & E News*, Apr. 25, 2021, online p 1, 2.
2. National Public Radio. "Marist National Poll: Extreme Weather." NPR/PBS NewsHour, Oct. 5, 2021.
3. Hauer, Mathew E. "Migration Induced by Sea-Level Rise Could Reshape the US Population Landscape." *Nature Climate Change*, Apr. 17, 2017, p 320.
4. Gleick, Peter. "The Climate Crisis Will Create Two Classes: Those Who Can Flee and Those Who Cannot." *The Guardian*, July 7, 2021, online p3.
5. Flavelle, Christopher. "Rising Seas Threaten an American Institution: The 30-Year Mortgage." *New York Times*, June 19, 2020, online p 3.
6. Hino, Miyuki and Marshall Burke. "The Effect of Information about Climate Risk on Property Values." *Proceedings of the National Academy of Sciences*, Apr. 27, 2021, online p 1.
7. Teirstein, Zoya. "Homes in Flood Zones Are Overvalued by Billions, Study Finds." *Grist*, Feb. 16, 2023, online.
8. Mercado, Angely. "Climate Gentrification Is Coming to Hurricane-Wrecked Florida." gizmodo.com, Jan. 13, 2023, online p 1.
9. Milman, Oliver. "'We're Moving to Higher Ground': America's Era of Climate Mass Migration Is Here." *The Guardian*, Sep. 24, 2018, p 3.

10. Pulkkinen, Levi. "Hamilton Might Be the Most Flooded Town in Washington." *Crosscut*, Feb. 4, 2020.

11. Pulkkinen, and interviews.

12. America Adapts, Climate Change Podcast. "Adapting to Chronic Flooding: Survivors' Stories and Actions They Take!" Climate Change Podcast, June 6, 2022.

13. Interview, 2022.

14. Schwartz, John. "As Floods Keep Coming, Cities Pay Residents to Move." *New York Times*, July 5, 2019, online p 10.

15. Bittle, Jake. "Climate Change Is Already Regiggering Where Americans Live." *The Atlantic*, Sep. 3, 2021, online p 2.

16. Teirstein, Zoya. "Florida Republicans Are Ready to Stop Rising Seas—Just Not Climate Change." *Grist*, Mar. 9, 2021.

17. Magill, Keith. "Move Where?" *Daily Comet*, Thibodeaux, Louisiana, May 30, 2015, online p 1.

18. Alig, Ralph J. et al. "Area Changes in U.S. Forests and Other Major Land Uses, 1982 to 2002." US Dept. of Agriculture, online p 3.

19. Revkin, Andrew. "Development and Disasters—A Deadly Combination Well Beyond Houston." *ProPublica*, Sep. 30, 2017, online p 3.

20. Diringer, Elliott. "Valley Floodwaters Recede—Threat of Inundation Does Not." *San Francisco Chronicle*, Mar. 3, 1997, A8.

21. Personal communication with Katharine Mach, 2021.

22. Dunagan, Christopher. "Floodplain Projects Open Doors to Fewer Floods and More Salmon." *Encyclopedia of Puget Sound*, Apr. 11, 2017.

23. National Institute of Building Sciences. *Natural Hazard Mitigation Saves*, 2018, p 1.

24. US Census Bueau. *Calculating Migration Expectancy Using ACS Data*. Apr. 11, 2022, online.

25. Godschalk, David R. et al. *Natural Hazard Mitigation*. Island Press, 1999, p 188.

26. Association of State Floodplain Managers. *Flood Mapping for the Nation*. Jan. 2020, p 3, derived from FEMA FloodSmart website.

27. Flavelle, Christopher. "Bank Regulators Present a Dire Warning of Financial Risks from Climate Change." *New York Times*, Oct. 17, 2019, online p 1.

28. Flavelle, Christopher. "Florida Sees Signals of a Climate-Driven Housing Crisis." *New York Times*, Oct. 12, 2020, online p 1.

29. Lustgarten, Abrahm. "Climate Change Will Force a New American Migration." *ProPublica*, Sep. 15, 2020, p 9, 15.

30. Jackson, Candice. "What Does It Take to Build a Disaster-Proof House?" *New York Times*, Nov. 12, 2021, p 3.

31. Schneider, Mike. "Survey: 3.3 Million US Adults Displaced by Natural Disasters." AP News, Jan. 5, 2023, online p 1.

32. Choi-Schagrin, Winston. "After the Storm, the Mold: Warming Is Worsening Another Costly Disaster." *New York Times,* Oct. 8, 2022.

33. Schwartz, John. "As Floods Keep Coming, Cities Pay Residents to Move," *New York Times,* July 5, 2019, online.

34. Morris, Frank. "Losses Continue to Mount After Recent Midwest Flooding." *Morning Edition,* National Public Radio, Apr. 22, 2019, online p 3,4.

35. Godschalk, p 4, 38, 45, 65.

36. Weber, Anna. *Going Under: Long Wait Times for Post-Flood Buyouts Leave Homeowners Underwater.* Natural Resources Defense Council, 2019, online p 1.

37. Mach, Katherine J. et al. "Managed Retreat Through Voluntary Buyouts of Floodprone Properties." *Science Advances,* Oct. 9, 2019, p 1.

38. Kann, Drew. "Millions More US Homes Are at Risk of Flooding Than Previously Known, New Analysis Shows." *CNN Weather,* June 29, 2020, online.

39. Moore, Rob. "As Climate Risks Worsen, U.S. Flood Buyouts Fail to Meet the Need." *Yale Environment 360,* Jan. 31, 2023, online p 2.

40. Colman, Zack. "Dream Homes and Disasters." *Politico,* Nov. 25, 2022, online p 3.

41. Bradstream, Lana. "Swanson Remembered for 'Unmatched' Vision." *Rapid City Journal,* June 5, 2009, online p 1.

42. Bureau of Outdoor Recreation. "Flood Plains for Open Space and Recreation." *Outdoor Recreation Action,* spring 1976, p 6.

43. Becker, William S. *The Creeks Will Rise.* Chicago Review Press, 2021, p 38, 143.

44. Merritt, Raymond H. *The Corps, the Environment, and the Upper Mississippi River Basin.* Army Corps of Engineers Historical Division, 1984, p 75–81.

45. Interview, 2021.

46. Faber, Scott. *On Borrowed Land.* Lincoln Institute, 1996, p 2.

47. Freitag, Bob et al. *Floodplain Management.* Island Press, 2009, p 117, 131.

48. Pinter, Nicholas. "The Long U.S. History of Relocating Communities Because of Flooding." *Slate,* Sep. 7, 2021, p 5.

49. Glick, Patty et al. *Natural Defenses from Hurricanes and Floods.* National Wildlife Federation, 2014, p 39. Also Anderson, Mariam Gradie and Rutherford H. Platt. "St. Charles County, Missouri: Federal Dollars and the 1993 Midwest Flood." In Mileti, *Disasters and Democracy,* 1999, p 215–238.

50. Faber, p 3.

51. Braun, Stephen. "Midwesterners Bailing Out of Floodprone River Plains." *Los Angeles Times,* July 14, 1995, A1, A7. Also Interagency Floodplain Management Review Committee. *Sharing the Challenge.* 1994, Table 1.4. In Mileti, *Disasters and Democracy,* p 233.

52. American Rivers. *In Harm's Way: The Costs of Floodplain Development,* July 1999, p 13, 18.

53. Interview, 2021.
54. Interview, 2021. Also Lohan, Tara. "Let Rivers Flood: Communities Adopt New Strategies for Resilience." *The Revelator,* July 8, 2019.
55. Interview, Roger Lindsey, 2022.
56. Hinshaw, p 199.
57. Wertz, Joe. "How Tulsa Became a Model for Preventing Floods." National Public Radio, Nov. 20, 2017, online p 2. Also American Rivers. *In Harm's Way: The Costs of Floodplain Development.* July 1999, p 17, 19.
58. FEMA. *Teamwork Approach to Outreach and Engagement Reduces Flood Risk.* Mar. 10, 2022, online.
59. Paulsen, Stephen. "Flood the Market." *Grist,* Aug. 27, 2019, online p 1-11.
60. Becker, p 123.
61. Associated Press. "Florida Sees Rise in Flesh-Eating Bacteria Amid Ian Concerns." AP, Oct. 19, 2022, online p 1.
62. Moore, Rob. "Seeking Higher Ground: Climate Smart Solutions to Flooding." Natural Resources Defense Council blog, July 25, 2017.
63. Interview, 2020.
64. Hunn, David et al. "Build, Flood, Rebuild: Flood Insurance's Expensive Cycle." *Houston Chronicle,* Dec. 7, 2017.
65. Walsh, Mary Williams. "A Broke, and Broken, Flood Insurance Program." *New York Times,* Nov. 4, 2017, online p 2, 4.
66. King, R. O. *National Flood Insurance Program: Background, Challenges, and Financial Status.* Congressional Research Service, 2009.
67. FEMA News Desk, personal communication, Mar. 11, 2022. Also FEMA. *Repetitive Loss Fact Sheet.*
68. Hersher, Rebecca. "Inspector General Slams FEMA Over Repeatedly Flooded Homes." National Public Radio, Sep. 15, 2020.
69. Hofflower, Hillary. "Eleven States Pay More in Federal Taxes Than They Get Back." *Business Insider,* Jan. 14, 2019, online p 1. Also Schultz, Laura. *Giving or Getting? New York's Balance of Payments with the Federal Government.* Rockefeller Institute of Government, Jan. 18, 2021.
70. Lightbody, Laura. "Repeatedly Flooded Properties Will Continue to Cost Taxpayers Billions of Dollars." pewtrusts.org, Oct. 10, 2020.
71. Hunn, p 4.
72. Hunn, p 7.
73. Flavelle, Christopher. "Canada Tries a Forceful Message for Flood Victims: Live Someplace Else." *New York Times,* Sep. 10, 2019, online p 1-5.
74. Jingan, Huo and Rebecca Hersher. "The Federal Government Sells Flood-Prone Homes to Often Unsuspecting Buyers, NPR Finds." National Public Radio, Sep. 14, 2021.
75. FEMA News Desk, personal communication, 2022.
76. Mileti, Dennis S. *Disasters by Design.* Joseph Henry Press, 1999, p7.

77. Flavelle, Christopher. "Billions for Climate Protection Fuel New Debate: Who Deserves It Most." *New York Times,* Dec. 3, 2021.

78. Government Accountability Office. *Disaster Recovery: Actions Needed to Improve the Federal Approach.* The Office, 2022.

79. Mervosh, Sarah. "'I Don't Want to Stay Here': Half a Million Live in Flood Zones, and the Government Is Paying." *New York Times,* Apr. 10, 2019, online p 1.

80. Flavelle, Christopher and Kalen Goodluck. "Dispossessed, Again: Climate Change Hits Native Americans Especially Hard."*New York Times,* June 27, 2021, online p 5.

81. Frank, Thomas. "FEMA Flood Program Could Violate Civil Rights Law." *Politico,* June 16, 2022, p 4.

82. Frank, Thomas. "Disaster Loans Foster Disparities in Black Communities." *E & E News,* June 25, 2020, online p 1–5.

83. Anthropocene Alliance. *The Great American Climate Migration.* Jan. 15, 2022, online p 2.

84. Flavelle, Christopher. "Why Does Disaster Aid Often Favor White People?" *New York Times,* June 7, 2021, online p 2, 3, 5.

85. Huang, Paul, State Association of Floodplain Managers annual conference, 2021.

86. Frank, Thomas. "FEMA Flood Program Could Violate Civil Rights Law." *Politico,* June 16, 2022, p 11.

87. Gleick, Peter. "The Climate Crisis Will Create Two Classes: Those Who Can Flee, and Those Who Cannot." *The Guardian,* July 7, 2021, online p 1.

88. Flavelle, Christopher. "In Houston, a Rash of Storms Tests the Limits of Coping with Climate Change." *New York Times,* Oct. 2, 2019, online p 4.

89. Flavelle, Christopher. "America's Last-Ditch Climate Strategy of Retreat Isn't Going So Well." *Bloomberg,* May 2, 2018, online p 3–14.

90. Klein, Naomi. "Why Texas Republicans Fear the Green New Deal." *New York Times,* Feb. 21, 2021, online.

91. Leopold, Luna and Thomas Maddock. *The Flood Control Controversy.* The Conservation Foundation, 1954, p 24.

92. Interview, 2022.

93. Shogren, Elizabeth. "Where FEMA Fails." *High Country News,* Aug. 31, 2015, online.

94. Interview, 2022.

95. Shapiro, Ari. "FEMA Coordinator Describes Catastrophic Flooding in Kentucky." *All Things Considered,* National Public Radio, Aug. 2, 2022, online p 4.

96. Maciag, Mike. "Building Homes in Flood Zones: Why Does This Bad Idea Keep Happening?" *Governing,* July 27, 2018.

97. Dorothy, Olivia. *The Multiple Benefits of Floodplain Easements.* American Rivers, Sep. 2022, p 8.

98. Moore, Rob. "As Climate Risks Worsen, U.S. Flood Buyouts Fail to Meet the Need." *Yale Environment 360*, Jan. 23, 2020, online p 1.

99. Flavelle, Christopher. "Why Ian May Push Florida Real Estate Out of Reach for All but the Super Rich." *New York Times*, Oct. 13, 2022, online p 6.

100. Goodell, Jeff. *The Water Will Come*. Little, Brown and Company, 2017, p 275.

101. Frank, Sadie et al. *Inviting Danger*. Brookings Institution, 2021, cited in *News & Views*, the Association of State Floodplain Managers newsletter, June 2021.

10. GREENWAYS

1. American Rivers. *Economic Impacts of Rivers and Recreation*. American Rivers, 2014.

2. Opperman, Jeffrey J. *Floodplains*. University of California Press, 2019, p 2.

3. Wallin, Phil. "From the President." *River Voices*, newsletter of River Network, summer 1997, p 3.

4. Palmer, Tim. *Great Rivers of the West Survey*. Western Rivers Conservancy, 2020.

5. Johnson, Kris. "Conserving Floodplains to Mitigate Future Flood Risk." *Nature Sustainability*, Dec. 9, 2019.

6. Serra-Llobet, Anna et al. "Restoring Rivers and Floodplains for Habitat and Flood Risk Reduction: Experiences in Multi-Benefit Floodplain Management from California and Germany." *Frontiers in Environmental Science*, Mar. 16, 2022, p 1.

7. National Park Service, Rivers and Trails Conservation Assistance. *Economic Impacts of Protecting Rivers, Trails, and Greenway Corridors*. National Park Service, 1990.

8. Power, Thomas Michael. *Lost Landscapes and Failed Economies*. Island Press, 1996, p 41.

9. Hise, Greg and William Deverell. *Eden by Design: The 1930 Olmsted-Bartholomew Plan for the Los Angeles Region*. University of California Press, 2000.

10. Amigos de los Rios. *Stewards of the Emerald Necklace Greenway*. Amigos de los Rios website, 2023.

11. Medberry, Mike. *Living in the Broken West*. Fish Hawk, 2022.

12. Stevens, W. K. "Restoring Wetlands Could Ease Threat of Mississippi Flood." *Science Times*, Aug. 8, 1995.

13. Faber, Scott. "Flood Policy and Management: A Post-Galloway Progress Report." *River Voices*, River Network newsletter, summer 1997, p 4.

14. Palmer, Tim. *America by Rivers*. Island Press, 1996, p 130.

15. Williams, Philip & Associates. *An Evaluation of Flood Management Benefits through Floodplain Restoration on the Willamette River, Oregon.* River Network, 1996, p 3.

16. Mileti, Dennis S. *Disasters by Design.* Joseph Henry Press, 1999, p 1, 6.

17. Glick, Patty et al. *Natural Defenses from Hurricanes and Floods.* National Wildlife Federation, 2014, p 5.

18. Saiyid, Amena H. and Kellie Lunney. "Wide-Ranging Water Infrastructure Bill Easily Passes House." Bloomberg Law News, July 30, 2020, online p 4.

19. Loller, Travis. "Corps of Engineers Considers Nature-Based Flood Control." AP News, Oct. 5, 2021.

20. Interview, 2022.

21. Sommer, Lauren et al. "Who Will Pay to Protect Tech Giants from Rising Seas?" National Public Radio, July 27, 2021, online.

22. Frazin, Rachel. "FEMA Restores Climate Consideration in Strategic Plan after Trump Dropped It." *The Hill*, Dec. 9, 2021.

23. Young, Robert S. "To Save America's Coasts, Don't Always Rebuild Them." *New York Times*, Oct. 4, 2022, online p 2.

24. Berginnis, Chad. "Historic Opportunity for Lasting Change." *News & Views*, Association of State Floodplain Managers, Dec. 2021, p 4.

25. FEMA News Desk, personal communication (no individual was named), Mar. 11, 2022.

26. FEMA News Desk, personal communication, Mar. 11, 2022.

27. Interview, 2023.

28. Interview, 2022.

29. Sherfinski, David. "After Ida, U.S. Cities Eye More Equal Resilience Plans." Reuters, Sep. 21, 2021, online.

11. LIVING WITH RIVERS

1. Romero, Ezra David. "Scientists Warn California's Floods May Be a Sample of the Megafloods to Come." *All Things Considered*, NPR, Apr. 6, 2023, archived online.

2. Intergovernmental Panel on Climate Change. "Climate Change Widespread, Rapid, and Intensifying." IPCC website, Aug. 9, 2021, online p 5.

3. Interview, 2021.

4. Tyson, Alec and Brian Kennedy. *Two-Thirds of Americans Think Government Should Do More on Climate.* Pew Research Center, June 23, 2020, online.

5. Plumer, Brad and Raymond Zhong. "Climate Change Is Harming the Planet Faster Than We Can Adapt, U.N. Warns." *New York Times*, Feb. 28, 2022, p 2.

6. Iglesias, Virginia et al. "Risky Development: Increasing Exposure to Natural Hazards in the United States." *Earth's Future*, June 2, 2021, online.

Index

Agnes Flood, 1–10, 33, 50, 107–108, 135, 162, 169, 259
Allegheny River, 76–77
American Planning Association, 153
American River, 87, 97–99, 245
American Rivers (organization), 35, 45, 119, 218, 234, 247
American Society of Civil Engineers, 43, 112
Andrus, Cecil, 42
Anthropocene Alliance, 235
ARkStorm, 122
Army Corps of Engineers, 32, 73–76, 83, 146, 180; cost of projects, 143; critique of development, 80–84; floodplain management recommendations, 29, 31, 80; levees, 94–95, 104; nonstructural projects, 148, 152–153, 157, 220, 253; speculative development, 78–80
Association of State Floodplain Managers, 69–70, 125, 151, 153, 178–181, 183, 188, 191–192, 219, 229, 232, 239, 254
Association of Structural Movers, 229
Atlas 14 reports, 124–125
atmospheric rivers, 17–20
Auburn Dam, 42, 90

Baker, Victor, 18, 88
Barron, Rita, 152

Barry, John, 176
Belt, Charles, 104
Berginnis, Chad, 191, 219, 239–240, 254
Biden administration, 44, 139, 150, 192, 235–236, 246, 254
Biggert-Waters Flood Insurance Reform Act, 190
Blackwelder, Brent, 83–85, 89
Bloomgren, Patricia, 153
Blumenauer, Earl, 232
Boise River, 86, 249–250
Bridges, Todd, 253–254
Brinkley, Douglas, 21
Brody, Sam, 129, 232
Brower, Lincoln, 56–57
Brown, Jerry, 136
Buffalo Creek Dam failure, 37
Bureau of Reclamation, 75, 86–87
Bush, George W., 175

Cain, John, 116, 256–257
California, 49; disasters in, 19; flood of 1862, 11, 14–15; levees in, 95, 97–99, 107, 109; open space protection, 246. *See also specific rivers, cities, and floods*
Canaan, Dave, 220–223
Canadian flood insurance policy, 232
Carter, Jimmy, 85, 148

305

306 INDEX

Cassatot River, 80
Center for Watershed Sciences, 90, 116
Centerville, IL, 234
Central Valley Flood Protection Plan, 116, 163
Central Valley of California. *See* Sacramento River; San Joaquin River
channelization, 105–106
Charles River, 152
Charlotte, NC, 208–215, 220–223, 230, 249
Chicago, 126
Clarion River, 89
climate change. *See* global warming
Clinton administration, 35, 154, 156, 175
coastal floodplains, 164, 173–174
Colorado River, 18, 39
Colton, Craig, 206
Columbia River, 105
Commoner, Barry, 102
Community Rating System of FEMA, 227
Connecticut River, 56–57
Cosumnes River, 101, 109, 117
Coyle, Kevin, 218, 236
Criss, Robert, 105
Criswell, Deanne, 127, 175
Cumberland River, 90
Curbelo, Carlos, 205

dams, 71–91; arguments against building more of, 89–90; costs of, 34, 89, 90, 143, 152; dams failing to contain floods, 85–88; dam failures causing floods, 25, 27–28, 33, 36–45, 90; flood control statistics, 43, 50; history of flood control dams, 72–78; removal of dams, 45; siltation of reservoirs, 91; ultimate fate of dams, 89–91. *See also* Army Corps of Engineers; *and specific dams and rivers*
debris flows, 25
Decker, Edwin, 104
Delaware River, 85
Dettinger, Michael, 14, 18
Diffenbaugh, Noah, 37
dikes. *See* levees
Diringer, Elliot, 115
disadvantaged populations, 102–103, 126, 233–236
Dolin, Eric Jay, 175
Dorian (hurricane), 48

earthquake dangers, 101
Edgar Jadwin Plan, 93

elevation of houses, 224
Ellet, Charles, 29
Ellicott City, MD, 124
Elwah River, 45
estuaries, 82

Faber, Scott, 35
Fahlund, Andrew, 163, 256
Fargo, ND, 118
farming and levees, 94
Feather River, 99, 116
Fedarko, Kevin, 39
Federal Emergency Management Agency, 45, 94, 114, 116, 125–129, 137, 152, 157, 160, 168, 170–192, 196, 214, 219–222, 229–234, 237–239, 253–255
Federal Interagency Floodplain Management Task Force, 89
fill-and-build, 177–179, 188
First Street Foundation, 126, 137–138, 188, 195, 216
Fish and Wildlife Coordination Act of 1934, 81
Flint River, 85
floodplain management, 141–165; origin of name, 147
floodplains: acreage of, 21, 60, 142, 152; converted to development, 164; defined, 60; natural values, 52–70; water storage, 69. *See also specific floodplain topics*
flood proofing, 225, 229–230
floods: beach formation, 68; costs of, 49, 135, 143; deaths from, 23, 27; disclosure of risks, 181–184; ecosystem benefits, 57, 61–69, 243–244; Flood Control Act of 1917, 29, 73, 93; Flood Disaster Protection Act of 1973, 169; flood of 1862, 11, 14–15, 122; flood of 1927, 73, 94; flood of 1936, 30–32, 74, 93; forests, 64; groundwater, 62; increasing height of crests, 121–140; mapping, 123–129, 139, 176–181, 221; natural functions, 52–70; relocation of development (*see* relocation); silt transport, 59; types of floods, 22–26, 30, 33; wetlands, 63. *See also specific floods, cities, rivers, and hurricanes*
Florence (hurricane), 36, 123, 160
Folsom Dam, 87, 98–99
Four Twenty Seven (organization), 135
Fowler, David, 151, 156, 205
Friant Dam, 86
Friends of the River (California), 40, 87

INDEX   307

Fugate, W. Craig, 125
Furniss, Mike, 70

Galloway, Gerald, 128, 153–158, 171, 250, 264
Garamendi, John, 119
Gleick, Peter, 122, 236
Glen Canyon Dam, 39
global warming, 65, 127, 130–139, 265
Goddard, James, 147, 158
Goodell, Jeff, 241
Great Depression, 74–75
Greenbrier River, 22–23
green infrastructure, 252–255
greenways, 243–258; costs and benefits, 245–250. *See also* nonstructural projects
Guerneville, CA, 138

Hadd, John, 145
Halpin, Eric, 44
Hamilton, WA, 197–204
Hannibal, MO, 104
Harp, Brian, 134
Harvey (hurricane), 35, 134–135
Hazard Mitigation and Relocation Assistance Act of 1993, 219
history of flooding, 11–52; deep past, 15–16; timeline for, 12–13. *See also specific floods*
Houston, TX, 128, 229, 231
Hoyt, William, 145
Huang, Xingying, 135
Humphreys, Andrew, 29
hurricanes, 24, 35. *See also hurricanes by name*

Ian (hurricane), 49, 136, 230
Ida (hurricane), 49
Indian tribes, 234; Quillayute, 234; Quinault, 234; Seneca, 76–77; Stillaguamish, 117
Ingram, B. Lynn, 14, 18
Inhofe, James, 130
insurance. *See* National Flood Insurance Program
Interagency Floodplain Management Review Committee, 35, 154
Intergovernmental Panel on Climate Change, 132, 134, 139, 263
international flood issues, 48, 144, 195

Jamestown, 93
Jarvis, Brooke, 140

Jeffres, Carson, 116
Johnson, Lyndon, 145–146
Johnstown flood, 27–28, 32–33, 50
Jones, Lucy, 122

Kalman, Naomi, 187
Katrina (hurricane), 35, 113, 135, 190
Kazmann, Raphael, 116
Keenan, Jesse, 233
Kelly, Robert, 103, 110
Kentucky flood of 2023, 150
Keys, Benjamin, 214
Keys, John, 39, 86
Kinzua Dam, 76–78
Klamath River, 45
Klein, Naomi, 121, 237–238
Knoxville, TN, 165
Knudson, Tom, 150
Kossin, James, 136
Kovach, Tim, 41
Kroening, David, 223, 230, 249
Kusler, Jon, 161

Land and Water Conservation Fund, 246
Larson, Larry, 178–180, 189
LaVelle, John, 176, 228
Leopold, Luna, 89, 159, 238
levees, 29, 92–120; aggravation of flows, 104–105; at American River, 97–99; Army Corps support, 93, 114; California locations, 95, 107; costs, 95, 100, 107, 110, 113–115, 117–118; environmental problems, 96; failure of, 34–35, 93–94, 99, 107–113; fairness issues, 102–103; farm vs. urban areas, 95, 114, 118; historic development, 93; improvements in, 115–117; Mississippi River locations, 92–93, 103–104, 111–113; rodent damage, 109; statistics, 94–95, 114, 118
Lightbody, Laura, 143
Lindsey, Roger, 164, 182, 224–225, 228, 237
Livingston, MT, 125
logjams and floods, 65–66
Los Angeles basin, 19, 33, 37, 59, 248
Lumber River, 36, 123
Lund, Jay, 89–90
Lustgarten, Abraham, 214

Mach, Katharine, 207
mapping. *See* floods: mapping
Matthew (hurricane), 12, 160
Maurstad, David, 217

308 INDEX

Mayorkas, Alejandro, 192
McGinty, Katie, 154–155
McHarg, Ian, 149–150, 225
McKenzie River, 52–55
McKibben, Bill, 127
McShane, John, 57, 151
Mecklenburg County. *See* Charlotte, NC
Medberry, Mike, 250
Meral, Jerry, 246
Miami River, 73
Mierzwa, Mike, 163, 251, 262
Milenti, Dennis, 115, 159, 161, 164
Milwaukee, WI, 124, 220
Minnesota River, 250
Mississippi Delta, 82–83
Mississippi River, 18, 28–30, 93, 128; flood of 1927, 30; flood of 1993, 34–35, 111, 153, 218–220; flood of 2019, 45–48, 88, 112; levees, 93, 105, 111–113
Missouri River, 81–82, 94, 104
mitigation, 239–240. *See also* relocation; flood proofing
Mokelumne River, 101
Moore, Jeremy and Dorothy, 152–153
Moore, Rob, 162, 183, 187, 189, 192, 216, 230, 237
Morgan, Arthur, 73, 77
Mount, Jeffrey, 44, 96

Napa, CA, 220
Nashville, TN, 48, 164, 179, 223–225, 245
National Academy of Sciences, 112
National Center for Atmospheric Research, 134–135
National Dam Safety Program Act of 1996, 43
National Flood Insurance Program, 114, 129, 161, 163, 166–193, 236; costs and deficits, 173, 185, 213, 240; disclosure of damage potential, 181–184; equity issues, 184–185; fill-and-build, 177–179; justification of, 189; lobbying against, 170–171, 183, 186; payments made, 167; repeated payouts, 188, 231–233; risk rating reforms, 192; warnings given, 168, 172
National Hazards Center, 144, 161, 233
National Inventory of Dams, 37, 43
National Park Service Rivers, Trails, and Conservation Assistance Program, 247
National Water Commission, 80, 147
National Wildlife Federation, 252
Native Americans. *See* Indian tribes

natural disaster statistics, 267
Natural Resources Committee of FDR, 31
Natural Resources Defense Council, 162, 183, 187, 230. *See also* Moore, Rob
natural solutions. *See* mitigation; nonstructural projects; floodplain management
Nature Conservancy, 117
Nelson, Gaylord, 153
New Orleans, LA, 93, 113
New River, 23
New York City floods, 49, 179, 257
Niobrara River, 88
NOAA, 124, 128, 132, 134, 136, 143, 164
nonstructural projects, 148, 152, 163, 168, 220. *See also* floodplain management

Offutt Air Force Base, 47
Ohio River, 32, 72–73
Opperman, Jeff, 255
organizations involved in floodplain management. *See* American Rivers; Association of State Floodplain Managers; National Wildlife Federation; Natural Resources Defense Council
Oroville Dam, 39–41, 44, 90

Pacific Northwest National Laboratory, 134
Pactola Dam, 86
Palco, Tom, 224
Parmesan, Camille, 267
Pasterick, Ed, 128
Peak, Lori, 233
parks, establishment of. *See* greenways
Pew Charitable Trusts, 143
Pilkey, Orrin, 196
Pinter, Nicholas, 41, 104–105, 113, 118, 162, 205, 219
Pittsburgh, PA, 20, 30–31, 76
Platt, Rutherford H., 32, 161, 172
pollution on floodplains, 159–160
Pomperaug River, 245
population displacement by floods, 133, 135
Pralle, Sarah, 126
precautionary principle, 129
projections of future floods, 18–19
Puyallup River, 208

racial issues, 233–236
rainfall increasing, 122, 124, 128, 134–136
Rapid City, SD, 86, 217
Red River (KY), 85
relocation of houses and businesses from floodplains, 194–241, 260; after 1993

INDEX   309

flood, 218–219; benefit-cost, 211, 219; case studies, 217–229; delays, 229, 237; federal funding, 216, 239–240; intact communities, 228; market factors, 196, 214, 223, 241; numbers at risk, 195, 215; reluctance to move, 197–207; support for moving, 207–229; tax revenue issues, 213–215,
repeated damage and taxpayer costs, 231–233
Reuss, Martin, 80
Rio Chama, 93
Rio Grande, 82
river morphology and floods, 58–70
River Network, 251
River Partners, 116
Rivers and Harbors Flood Control Act of 1966, 147
Robinson, Bill, 227
Rogue River, 181
Roosevelt, Franklin, 31–32, 75, 189
Roosevelt, Theodore, 147
Russian River, 48

Sacramento Area Flood Control Agency, 99
Sacramento levees, 90, 93, 97–100, 108
Sacramento River, 29, 76; flood of 1986, 35, 95–99, 108–109; flood of 1996, 35; levees, 93, 99–100, 109–110
Sacramento-San Joaquin Delta, 110
salmon, 82, 105
Sandy (hurricane), 35
San Joaquin River, 42, 86, 100–101, 128, 256
Santa Cruz River, 128
Saylor, John, 77
Schad, Ted, 76
Schleicher, David, 38
sediment in reservoirs, 81–82
Severe Repetitive Loss Properties. *See* repeated damage
Shader, Eileen, 119, 234, 247
Shepherd, Marshall, 129
Shoji, Crystal, 238
Showstack, Randy, 218
Sidney, NY, 237
Skagit River, 197–204
Small Business Administration, 234
Soldiers Grove, WI, 217–218, 228
Spellmon, Scott, 157
Stanislaus River, 83–84
Steinbeck, John, 20, 205

St. Francis Dam failure, 37
Stillaguamish River, 117
St. Louis, MO, 104–105, 112, 117
Stockton, CA, 101
Stork, Ron, 42–43
Susquehanna River, 26, 237; Greenway Partnership, 260–261
Swain, Daniel, 18, 135, 137, 263
Swanson, Leonard, 217

Teton Dam failure, 37–38
Thompson, Carol, 208–211
Tittabawassee River dam failures, 37
Touma, Danielle, 127
Trinity Dam, 42
Tulare Lake, 104, 136, 262
Tulsa, OK, 225, 227
Tuolumne River, 85–86, 256
TVA, 75, 146–147, 158
Twain, Mark, 112
Tygart River, 73

U.S. Global Change Research Program, 134

Vileisis, Ann, 50–51
Vogel, Nancy, 150

Wallin, Phil, 245
Walls, Jerry, 260
Wanless, Harold, 132
Washington Department of Ecology, 257
Water Foundation, 162
Water Resources Council, 148, 155
Water Resources Development Act of 2020, 252
watershed connections, 126–127
Wesselman, Eric, 42
Western Rivers Conservancy, 245
White, Gilbert, 94, 144–145, 147, 149, 151, 154, 164, 168, 171, 188
wildlife and floods, 66–68
Willamette River, 251
Williams, Philip, 87, 251
wing dams, 105
Witt, James Lee, 175, 219
Wright, James, 191
Wright, Roy, 173

Young, Robert, 254

zoning, 158–162, 244; *Lucas v. South Carolina*, 161, 174

About the Author

PHOTO BY ANN VILEISIS

TIM PALMER has written thirty-two books about rivers, the environment, and adventure travel. Trained in landscape architecture and working as a county land-use planner, his involvement with floods dates to Hurricane Agnes in 1972, which he experienced firsthand as a flood victim and then in emergency response, mapping of floodplains, and establishment of zoning ordinances and floodplain acquisition programs. He continued to follow flooding issues nationwide as a planner, writer, photographer, and river conservation advocate through five decades that followed.

Among other honors, the author received the Conservation Achievement Award as Communicator of the Year from the National Wildlife Federation, the Lifetime Achievement Award from American Rivers, the Wild and Scenic Rivers Award from the River Management Society, the Distinguished Alumni Award from Penn State University College of Arts and Architecture, and the Ansel Adams Award for Photography from the Sierra Club. His books of text and photos have garnered the National Outdoor Book Award, Independent Publisher Book Award, IndieFab Adventure Book of the Year, IBPA Benjamin Franklin Award, and others.

Tim has written extensively about American rivers, conservation history, and natural resource policy, including frequent guest editorials in major West Coast newspapers. He lives on the coast of southern Oregon and speaks widely to academic, professional, conservation, and recreation audiences nationwide about rivers, flooding, forests, and adventure travel. See www.timpalmer.org.

Founded in 1893,
UNIVERSITY OF CALIFORNIA PRESS
publishes bold, progressive books and journals
on topics in the arts, humanities, social sciences,
and natural sciences—with a focus on social
justice issues—that inspire thought and action
among readers worldwide.

The UC PRESS FOUNDATION
raises funds to uphold the press's vital role
as an independent, nonprofit publisher, and
receives philanthropic support from a wide
range of individuals and institutions—and from
committed readers like you. To learn more, visit
ucpress.edu/supportus.